Second Edition

GENERAL CHEMISTRY IN ACTION

LABORATORY MANUAL
WITH
QUALITATIVE ANALYSIS

Na + Water---> Hydrogen Gas + NaOH, Heat

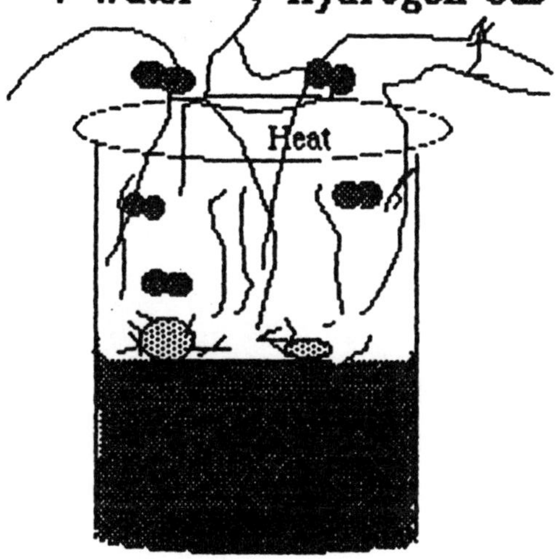

Johnnie J. Jenkins, Jr.
Ashok K. Kabi Satpathy
Ruhullah Massoudi
Frank Weaver
Andrew Koli
N. Datta-Gupta

KENDALL/HUNT PUBLISHING COMPANY
4050 Westmark Drive Dubuque, Iowa 52002

This edition has been printed directly from camera-ready copy.

ISBN 978-0-7872-5102-4

Printed in the United States of America
13 14 15 16 17 18 19 20

CONTENTS

EXPERIMENTS

APPENDICES

PREFACE

This laboratory manual is designed to accompany any of the General Chemistry or College Chemistry Texts. The experiments in this manual reinforce the learning of chemistry. The manual includes some experiments that have been typically provided in many manuals, however where possible, we have tried to be innovative in our approach. This lab manual is being designed in specific response to our observations and interactions over a six year period. Each of the experiments has general and specific objectives to assist the learning process. The experiments each have a Pre-Laboratory Assignment that must be completed prior to coming to the laboratory. They will be completed in the Chemistry Computer Center. There are a number of useful information bases included in the Appendices. The Pre-Lab Assignments will require the kinds of calculations and concepts that are presented/required in the experiments. The chemistry experience is enhanced by a solid laboratory component. The lab experience embodies the following:

1. Making chemistry come alive.
2. Providing an enjoyable lab experience.
3. Challenging the student's intellectual ability.
4. The experiments are relevant and illustrate important concepts.
5. Providing the student with a deeper appreciation and awareness of chemistry.
6. Enhancing the understanding of chemistry.

When working in the laboratory, you should always work independently unless instructed otherwise. Learn to be a self-achiever. In several of the experiments, you will be grouped for data collection purposes only. Each student is expected to complete and submit their individual laboratory reports. Don't waste time in the laboratory. To receive maximum benefit from the laboratory, you must be diligent, resourceful, and dedicated in your efforts. Always pay attention and read experiments before coming to the laboratory. The data and report sheet will be due one week upon completion of each experiment at the discretion of your laboratory instructor. The lab report will often include a report sheet/s from the manual and a sheet of calculations. All such calculation sheets or additional questions are to be completed on lineless paper. The report sheet should be stapled. Where calculations are to be done, please do them neatly so that they can be followed readily. Become familiar with all the laboratory regulations and abide by them. Always practice safety when conducting all lab experiments.

All experimental data should be written on the report sheets, and not on scraps of paper. **If an unknown is given, please remember to record it on the report sheet.** It is advisable that you obtain a pack of carbon paper. Place a sheet under the data sheet and put another sheet of plain paper under that for the purpose of having a second copy of the raw data to be collected. The Pre-lab assignments will be completed on the computer.

ACKNOWLEDGEMENTS:

We very gratefully thank the students for their suggestions, additions, deletions, and improvements in the development of this laboratory manual. A special thanks is extended to Mr. William (Bill) W. Wannamaker, P.E. for his candid review and recommendations. Special thanks is extended to Burgess Publishing Company for the use of several figures.

SAFETY RULES, REGULATIONS AND PROCEDURES

The following safety rules and procedures are mandatory for all students enrolled in chemistry laboratories. Noncompliance with these rules can result in dismissal from the laboratory. These rules are designed to ensure that maximum safety practices are adhered to in the laboratory. Your laboratory instructor should demonstrate the use of and point out the location of various safety equipment. The laboratory has fire extinguishers and fire blankets. Also, there is available a first aid kit located in the laboratory. Additionally, all injuries should be reported to your instructor. The appropriate measures will be undertaken in the event of an emergency or accident. Both major and minor cuts, bruises, burns and scratches must be reported immediately.

1. All students are required to wear safety glasses with side shields at all times while in the laboratory. The protection of the student is of primary concern in the laboratory. It is thus imperative that eye protection be worn to minimize serious accidents. This is a most important rule!

2. If a chemical gets into the eye, it is very important to flush it with copious quantities of water. A physician or the University Infirmary will be consulted.

3. No smoking, eating, drinking, or chewing are allowed in the laboratory.

4. If your hair or clothing catches fire, the fire should be smothered with a blanket or water except for a sodium or potassium fire. Remove the clothing if possible, and report to the College Infirmary or hospital.
   Gas burners should be immediately turned off.

5. Sitting on the laboratory bench tops is not allowed.

6. Never sniff on a chemical to detect an odor. The chemical should be placed in a suitable container and the vapors are wafted toward the face. The vapors of a number of chemicals are very potent.

7. When heating a substance in a test tube always use a Pyrex or Kimax brand test tube. Never point or aim a test tube at anyone while heating it.

8. Never taste a laboratory chemical.

9. You are expected to maintain a clean work environment. Do not put solids in the sinks. Prior to leaving the laboratory, please clean the sink and bench space.

10. Always wash your hands before leaving the laboratory.

11. Horseplay and playful activities are not allowed in the laboratory.

SAFETY RULES, REGULATIONS AND PROCEDURES CONTINUED

12. Please do not have visitors, friends coming to converse with you during the lab period.

13. Please follow the instructions of the laboratory instructor. Do not substitute chemicals or change the quantities of chemicals unless directed by your instructor.

14. Do not return liquid or solid reagents to bottles unless directed to by the instructor.

15. Do not leave the laboratory without informing the instructor.

16. Never perform unauthorized laboratory experiments.

17. Never work alone in the laboratory.

18. Reactions involving poisonous or irritating gases should be performed under a hood.

19. The handling of hazardous chemicals (corrosive acids, bases and other organic and inorganics) shall be under the supervision of the lab instructor. The disposal of such shall be directed by the instructor. Suitable containers for such are available in the laboratory.

20. All spills should be reported to the laboratory instructor. To neutralize spilled acid on clothing, add a mild solution of sodium bicarbonate of soda (baking soda), for spilled bases on clothing, add a boric acid or weak acid solution. For desktop acid or base spills add solid bicarbonate of soda.

21. When inserting glass tubing into rubber or cork stoppers, care should be taken to protect your hands. The end of the glass tubing should be fire polished, and a lubricant (water, if appropriate, mineral oil or silicon grease) applied to the end of the glass. The hands should be covered by gloves or lab cloth.

22. Never place hot objects on the balance pan when weighing.

23. Under no circumstances should chemicals be handled with bare hands.

24. Do not handle handle hot objects (glassware, etc.) with bare hands. Exercise caution when heating glass tubing and remember that it is hot.

25. It is strongly recommended that no sandals, shorts or other loosely hanging apparel be worn in the laboratory.

SAFETY RULES, REGULATIONS AND PROCEDURES CONTINUED

26. We strongly recommend that you minimize conversation so that you will be able to hear all instructions and any changes in them.
27. Wash all dirty glassware before leaving the laboratory at the end of each period.
28. Do not place solids, such as metals, paper, etc. into sinks.
29. Read all labels twice before obtaining and dispensing reagents.
30. Avoid obtaining more reagents than needed.
31. Never insert your dropper into a reagent bottle. This will avoid contamination.
32. Do not take reagent bottles to your desk, thus inconveniencing others.
33. Do not lay a stopper on the desktop , hold it between your fingers as you dispense.

WASTE DISPOSAL

Each of the laboratories has an inorganic and an organic waste disposal container. All of our laboratory experiments are being designed so as to minimize any hazardous waste which will help alleviate the problem of waste disposal. Please follow these guidelines as you dispose of chemicals.

1. Discard solids as directed by the lab instructor in appropriate labelled containers.
2. Place all organic wastes in designated containers.
3. Wash water soluble liquids into the drains with plenty of water.
4. Copper, silver, and mercury and heavy metal salts are very corrosive and hazardous, and must be disposed of in designated containers.

Campus Security : 536-7188 / 8811 (Dispatcher)

Poison Control Center : 1-800-922-1117

Emergency Medical Service : 803-531-3020 (Ambulance)

Brooks Health Center Infirmary : 536 - 7053/7055

BASIC LABORATORY EQUIPMENT AND SUPPLIES

DRAWER SUPPLIES:

1. Evaporating Dish (1)
2. Triangular File (1)
3. Watch Glass (1)
4. 10ml Graduated Cylinder (1)
5. 50ml Graduated Cylinder (1)
6. Short Stem Funnel (1)
7. 125 Erlenmeyer Flask (1)
8. Test Tube Holder(1)
9. Crucible and Cover (1)
10. Hard Glass Test tube (1)
11. Gas Lighter (1)
12. Wing Top (1)
13. Test Tubes 8 (13mm)
14. 150 ml Beaker (1)
15. 250 ml Beaker (2)
16. 400 ml Beaker (1)
17. Stirring rod (1)
18. Spatula (1)
19. Florence Flask 250ml (1)
20. Florence Flask 500ml (1)
21. Triangle Pipestem (1)
22. Thistle Tube (1)
23. Tongs (1)
24. Test tube Brush
25. Wire Gauze(1)
26. Stoppers
27. Bunsen Burner (1)
28. Pinch Clamp (1)
29. Cloth
30. Droppers (2)
31. Deflagrating Spoon

CABINET SUPPLIES:

1c. Litmus Paper
2c Mortar and Pestle
3c Test Tube Rack
4c Gas Collecting Bottles
5c Pneumatic Trough
6c Iron Ring
7c Ring Stand
8c Test Tube Rack
9c Buret Clamp
10c Glass Plates
11c Three Prong Clamps (2)

OTHER LAB SUPPLIES:

32. Safety Glasses
33. Electronic Balance
34. Buret
35. Pipet
36. Thermometer
37. Timer
38. Volumetric Flask
39. Reagent Bottle
40. Wash Bottle
41. Suction Flask
42. Transfer Pipet

The Laboratory Equipment and Supply Items are shown in Fig. 1.

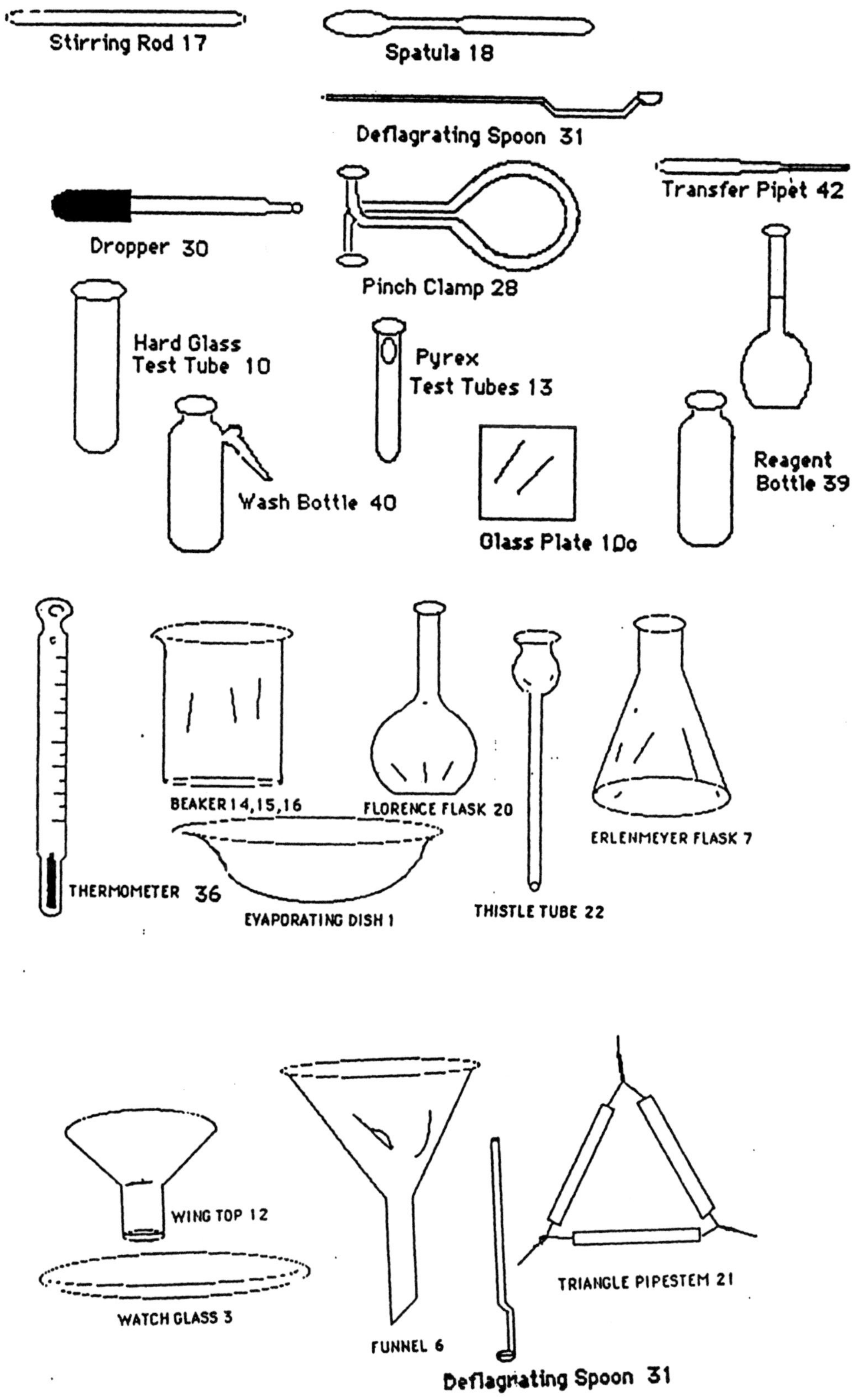

Fig 1. Common Laboratory Equipment and Supplies

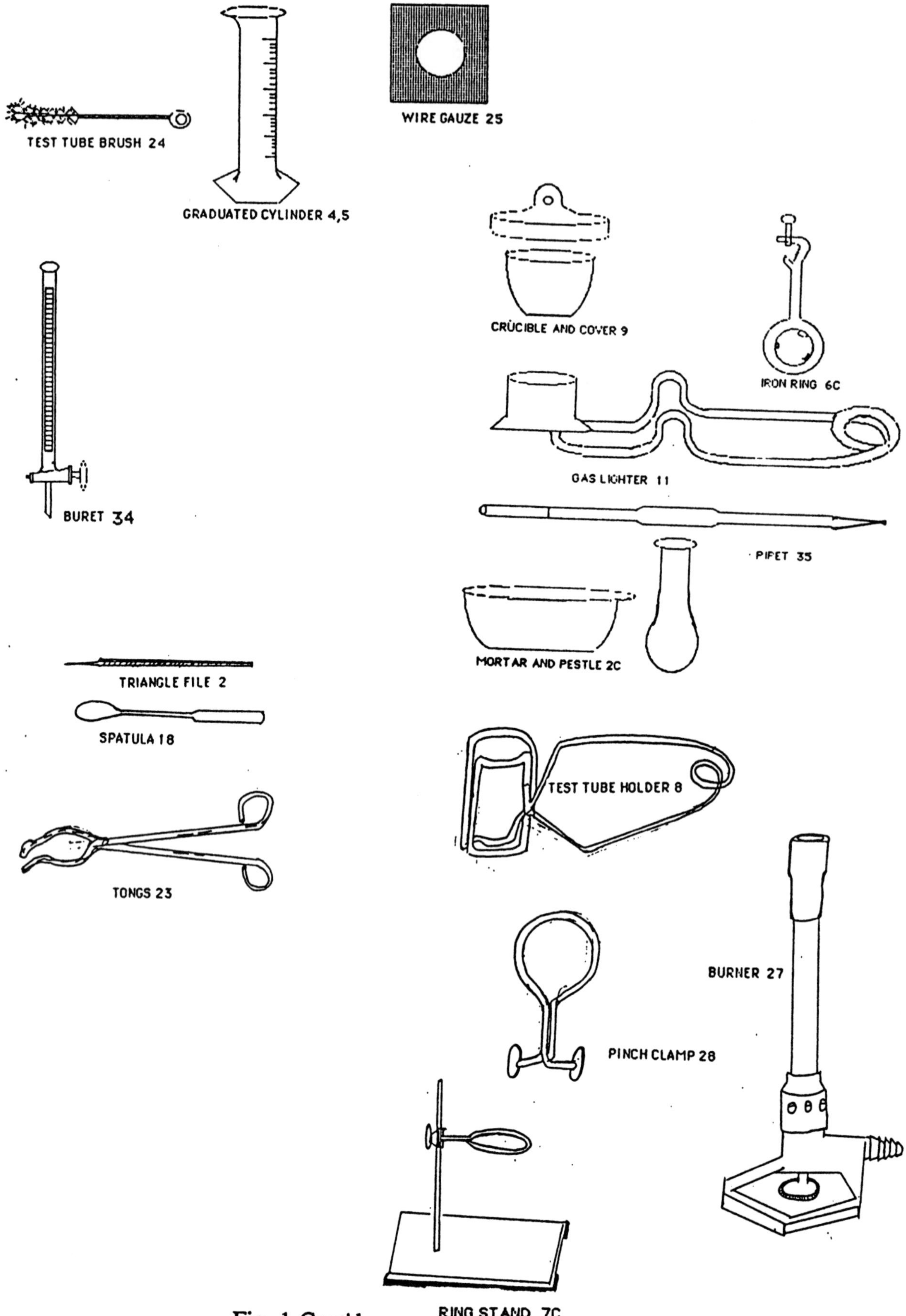
TEST TUBE BRUSH 24
GRADUATED CYLINDER 4,5
WIRE GAUZE 25
CRUCIBLE AND COVER 9
IRON RING 6C
GAS LIGHTER 11
BURET 34
PIPET 35
MORTAR AND PESTLE 2C
TRIANGLE FILE 2
SPATULA 18
TEST TUBE HOLDER 8
TONGS 23
BURNER 27
PINCH CLAMP 28
RING STAND 7C

Fig. 1 Cont'd

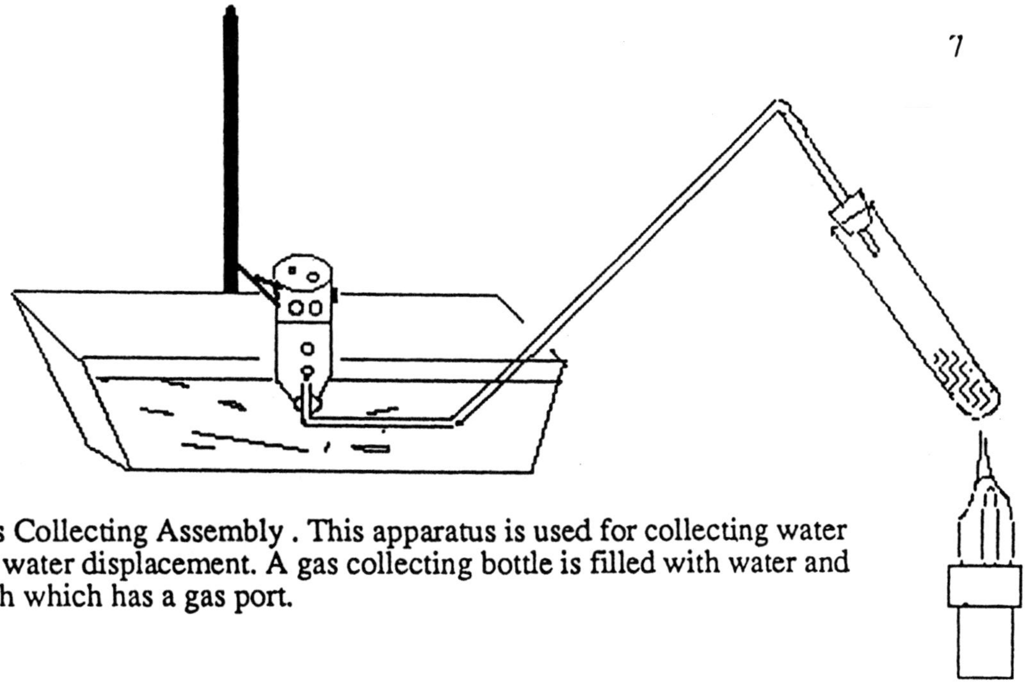

Technique 1T. Gas Collecting Assembly . This apparatus is used for collecting water insoluble gases by water displacement. A gas collecting bottle is filled with water and inverted in a trough which has a gas port.

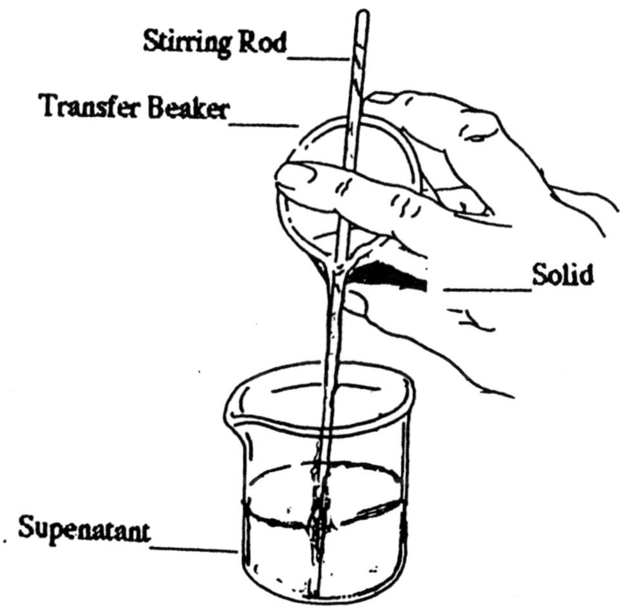

Technique 2T .Decanting a Liquid (supernatant) from a Solid . The liquid is poured slowly down a stirring rod with the beaker spout touching the stirring rod. Pouring it slowly minimizes solid (precipitate) mixing.

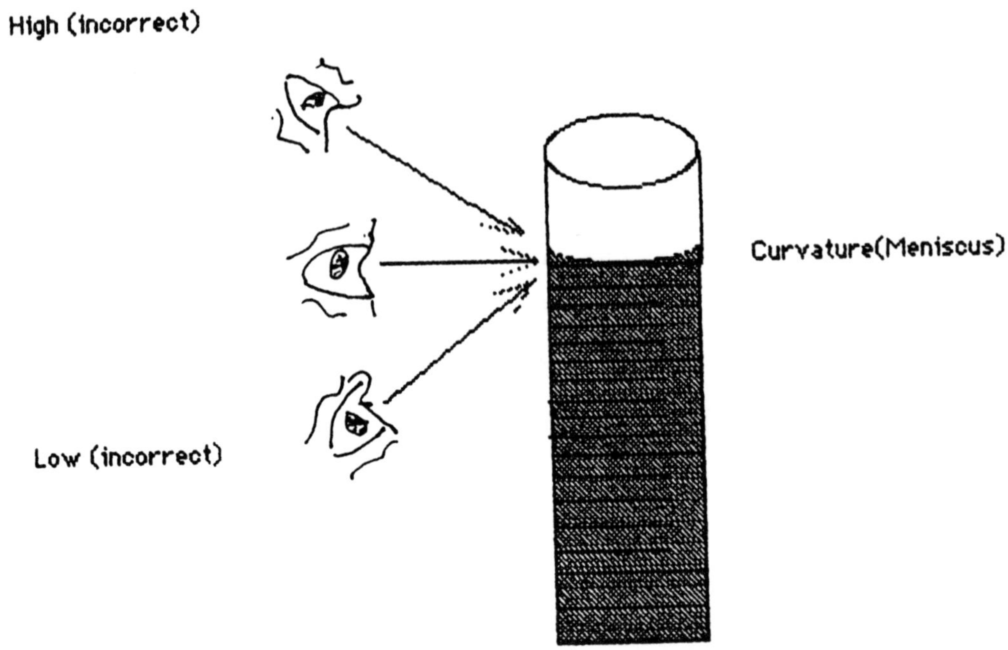

Technique 3T. Reading a Volume from a Graduated Cylinder. The proper sight is to read from the "bottom of the meniscus" (curvature) for wet liquids and at the top for non-wet .

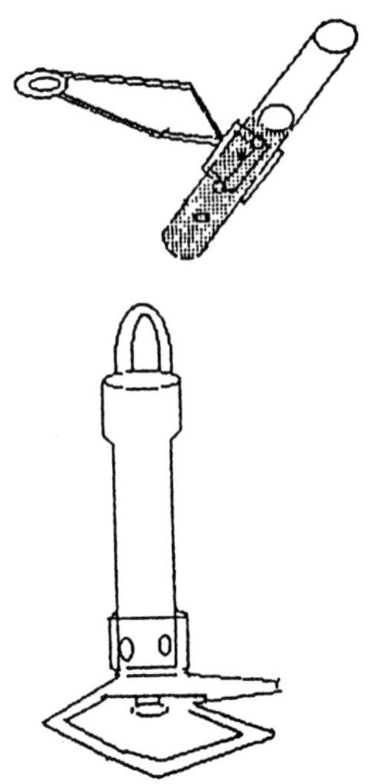

Technique 4T. Heating a Liquid in a Test Tube. The test tube is placed in the flame at approximately 45°. The test tube should be moved from top to bottom and in and out of the flame. The test tube should never be pointed at anyone. The test tube should be less than half full.

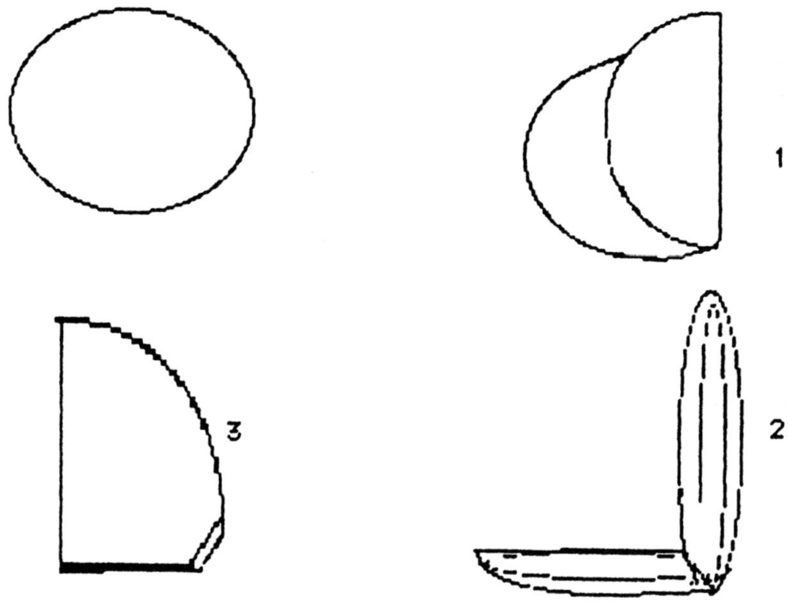

Technique 5T. Folding Filter Paper for Gravity Filtration. The paper is folded in half (1), then folded again (2), and a corner is torn off to allow it to fit the funnel (3).

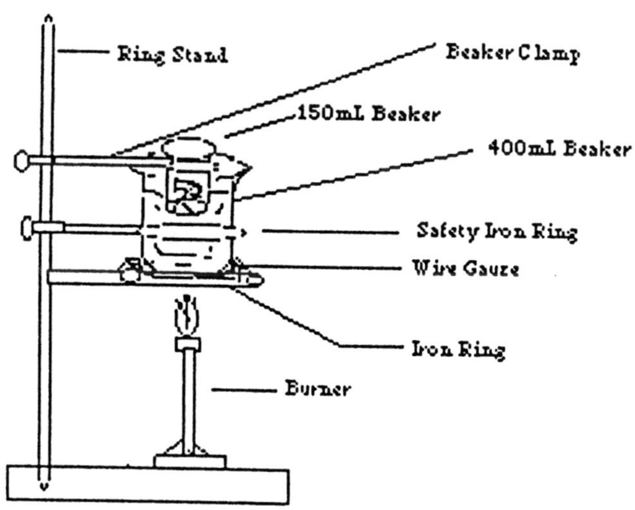

Technique 6T. Heating a Liquid in a Hot Water Bath. A liquid is placed in a beaker, and the beaker mounted in a hot water bath, and heated to the desired temperature.

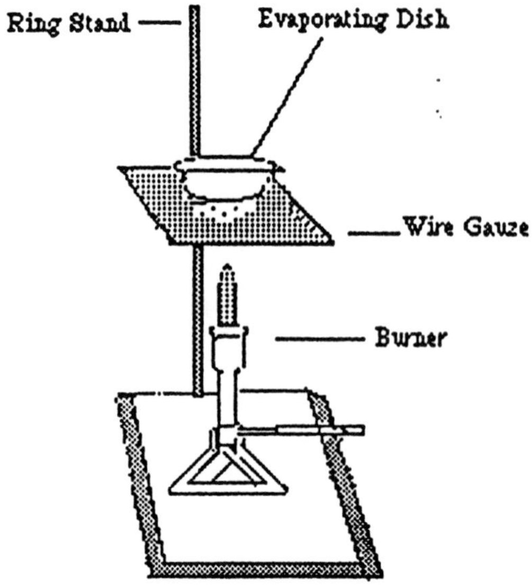

Technique 7T. Evaporating a nonflammable liquid. A liquid is placed in the evaporating dish and heated with a low (cool) flame.

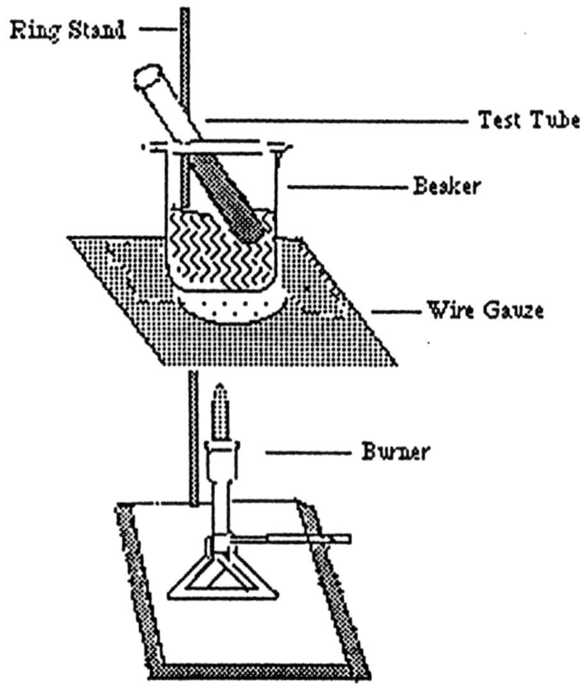

Technique 8T. Heating a Volatile Liquid in a Water Bath. The volatile liquid is placed in a test tube and placed in a water bath, and heated with a bunsen flame.

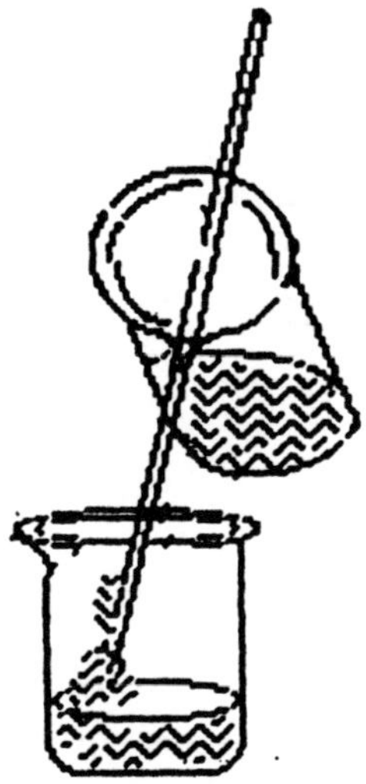

Technique 9T. Transferring a Liquid. The transfer beaker has a stirring rod touching the beaker spout while the liquid is poured into the receiving beaker. The stirring rod also touches the side of the receiving beaker.

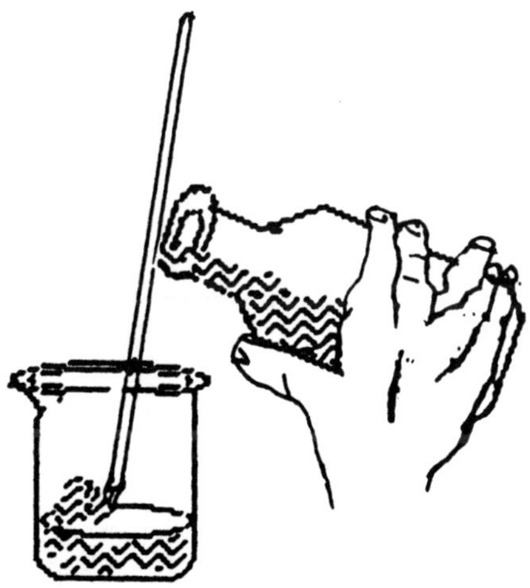

Technique 10T. Transferring a Liquid from a Bottle. The liquid is poured down a stirring rod into the receiving beaker. The stirring rod also touches the side of the receiving beaker.

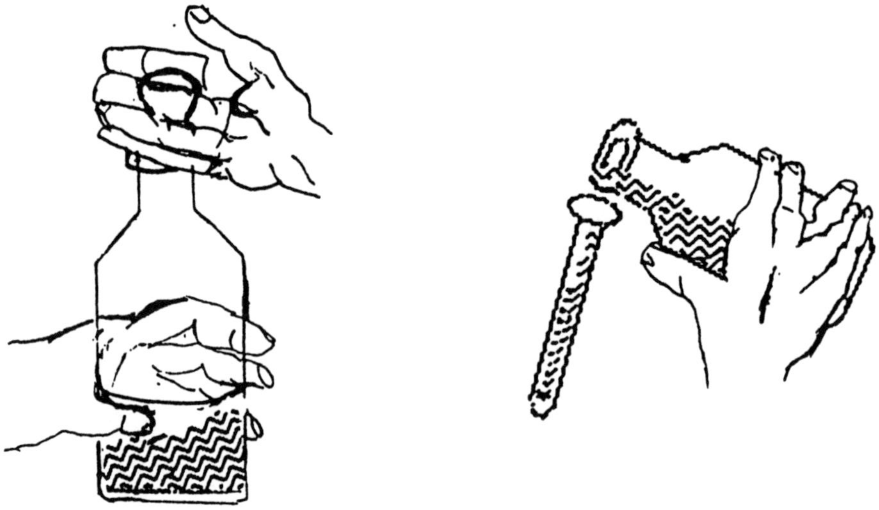

Technique 11T. Removing a Glass Stopper. The glass bottle is held while the stopper is removed. The stopper is held between the fingers of the opposite hand. Never lay the stopper on the laboratory bench.

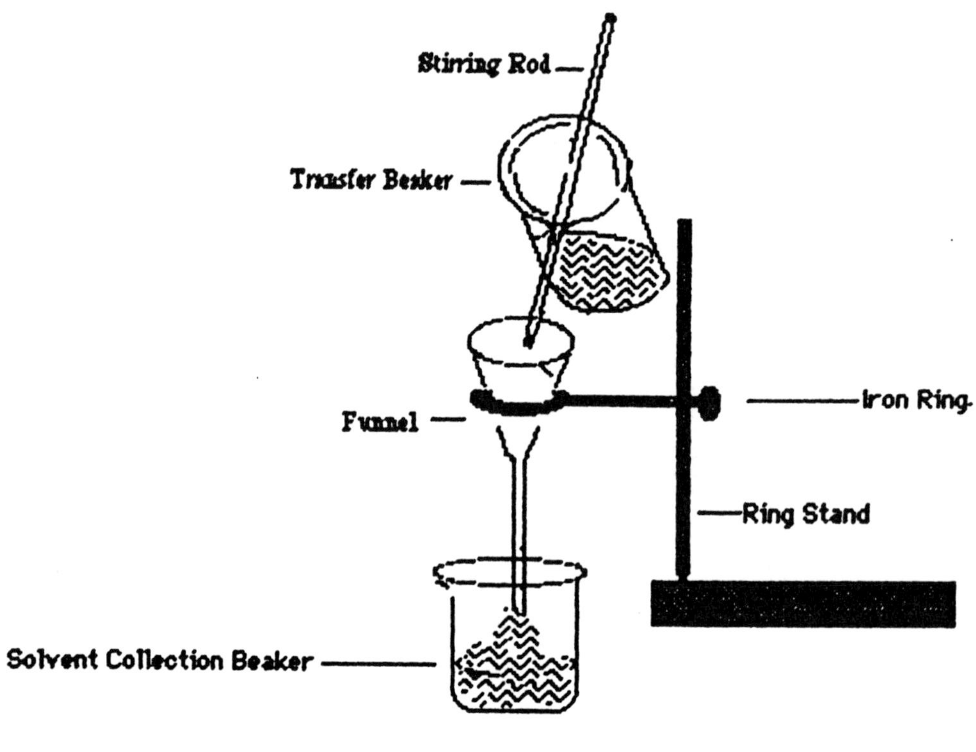

Technique 12T. Gravity Filtration. The folded filter paper is placed in the funnel and moistened with the solvent of the mixture. The solvent will be distilled water in most situations. Press the filter paper down to make a seal with the funnel. The mixture is poured down a stirring rod inserted against the wall of the filter paper. The tip of the transfer beaker should touch the stirring rod while the liquid is being poured. The funnel bowl should be filled less than 2/3 full. The mixture should never be poured above the filter paper line. The filtered solvent is collected in a beaker.

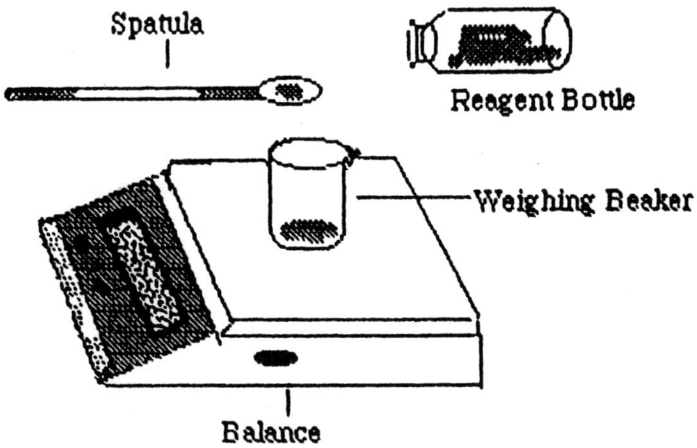

Technique 13T. Weighing a Powdered Solid. The solid is removed from the reagent bottle using a spatula or is poured while rotating the reagent bottle into a suitable weighing boat, beaker or other appropriate container. Solids are never poured directly onto the balance pan. The weight of the solid can be determined either by (1) weighing the beaker , recording its weight and adding desired solid until target weight is obtained, balance displays weight of beaker plus sample or (2) by taring the balance (re-zero) and add until target weight is obtained. The weight displayed is the weight of added solid.

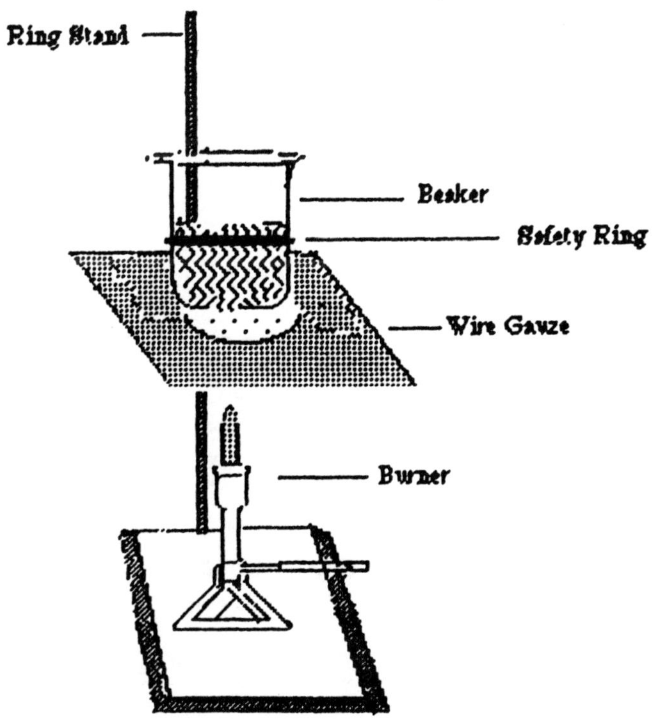

Technique 14T. A Hot Water Bath. The hot water bath has a beaker on a wire gauze that is approximately 1-1.5 inches above a bunsen burner. Their is an iron safety ring placed midway of the beaker.

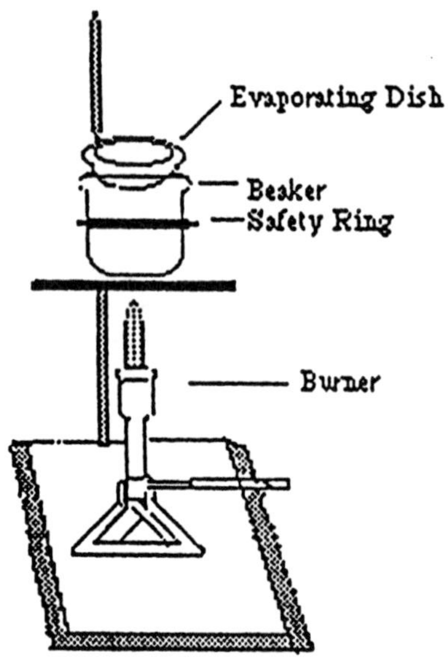

Technique 15T. The Evaporation of a Nonflammable Liquid Using a Steam Bath. The liquid is poured carefully into an evaporating dish. The dish is placed atop a beaker containing water which is to be heated to boiling.

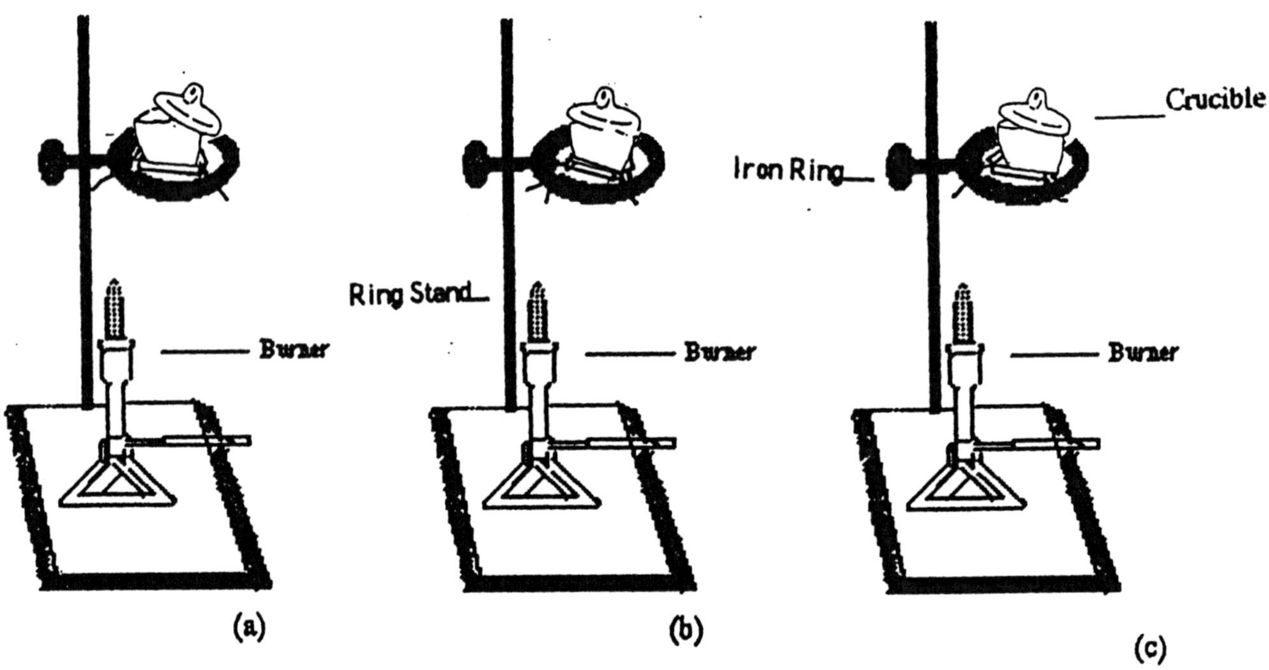

Technique 16T (a) . Drying and Firing a Crucible and Cover. The crucible and cover are placed on a triangle pipestem on an iron ring stand using the tongs. The ignition of a crucible without air mixing (b). The ignition of the crucible contents in the presence of air (c). Do not touch the crucible with hands after firing.

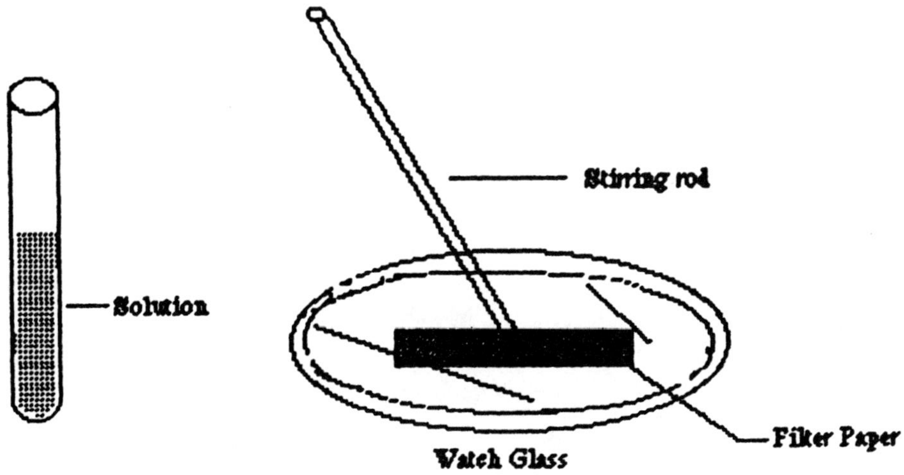

Technique 17T. Testing a Solution For Acidity. The stirring rod is inserted into the liquid and the rod is touched onto a piece of red and blue litmus paper on a watch glass.

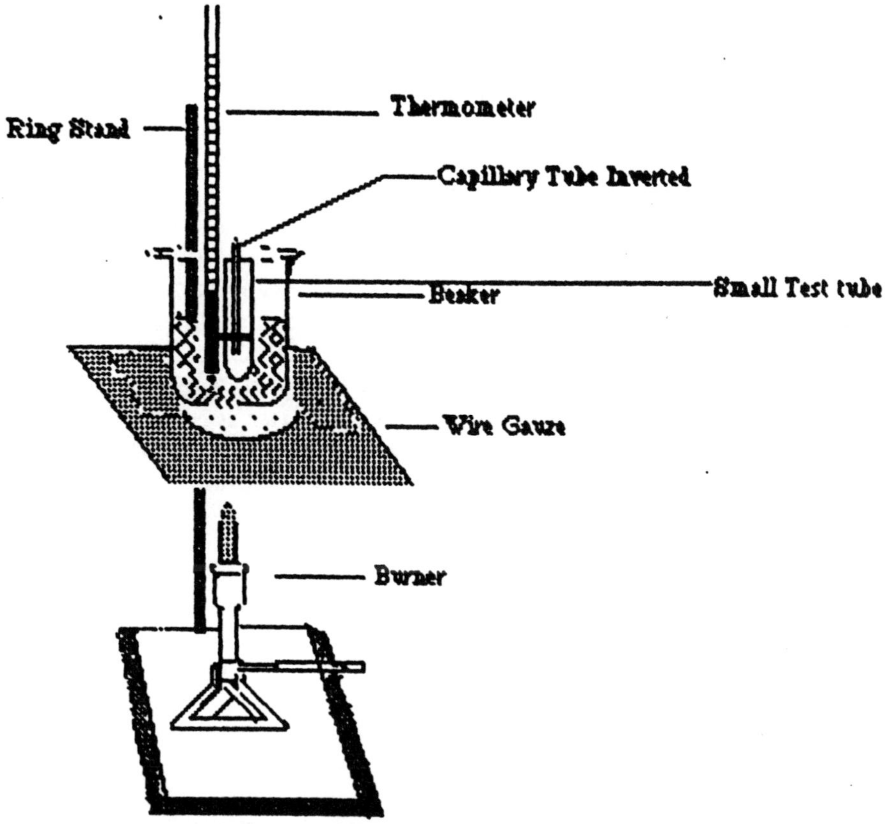

Technique 18T. Boiling Point Apparatus. A 150 ml beaker is placed on a wire gauze approximately half full of water or other appropriate solvent. Along side a thermometer is attached a small test tube. To the test tube is added 10 drops of liquid to be analyzed. The thermometer with test tube is mounted onto a ring stand using a three prong clamp. The thermometer is placed in the beaker with the test tube top above the solvent. A boileezer is placed in the test tube. An inverted capillary tube is inserted into the small test tube. The solvent is heated with a low "cool" flame. When bubbles rapidly appear at the mouth of the capillary tube, turn the heat off, and watch when the bubbles just cease and before any liquid enters the tube, read the thermometer and record this as the boiling point.

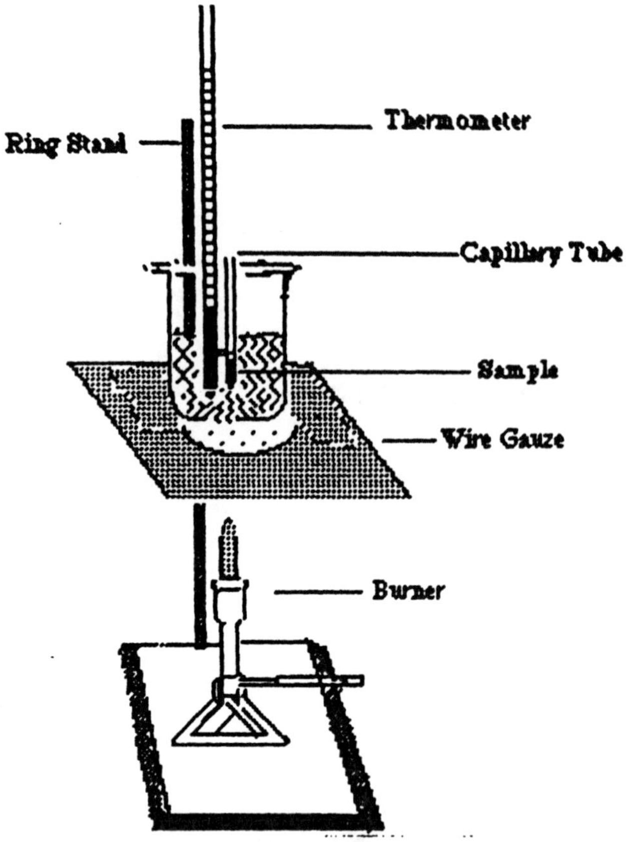

Technique 19T. Melting Point Apparatus. A 150 ml beaker is placed on a wire gauze approximately half full of water or other appropriate solvent. A small portion of the solid to be analyzed is placed in a mortar and pestle and ground to a powder. The open end of a capillary tube is inserted into the solid. A small amount of solid enters the tube. Using a glass rod approximately 1/2 meter in length, the capillary tube is dropped open end up down the glass tubing to pack the solid sample. The tube capillary tube containing solid is attached along side a thermometer with the bottom even with the thermometer bulb. The amount of sample need only be enough for visible detection. The thermometer with test tube is mounted onto a ring stand using a three prong clamp. The thermometer is placed in the beaker with the test tube top above the solvent. The solvent is heated with a low "cool" flame. You must carefully watch the thermometer and record the melting point. Note the temperature when crystals first appear to melt and when they have all finished and the first meniscus.

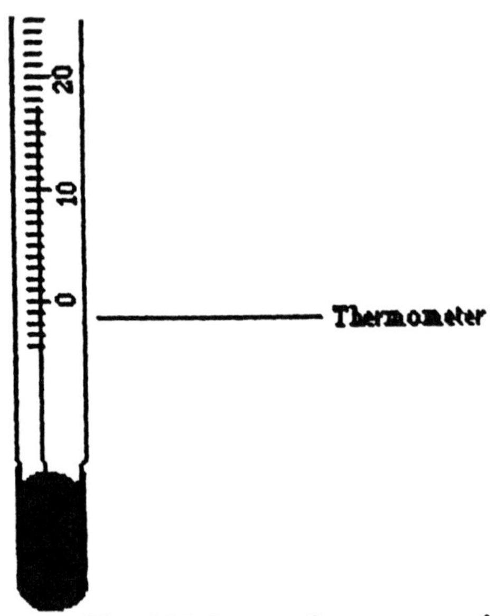

Technique 20T. Reading a Thermometer. The 250 degree thermometer is graduated in 1 degree increments. If the mercury head is between the degree lines, an estimate is made for the **0.1** place.

Technique 21T. Detecting Chemical Odors. The liquids are placed in an evaporating dish and the vapors are wafted toward the face for odor detection. Never should you sniff up on a chemical.

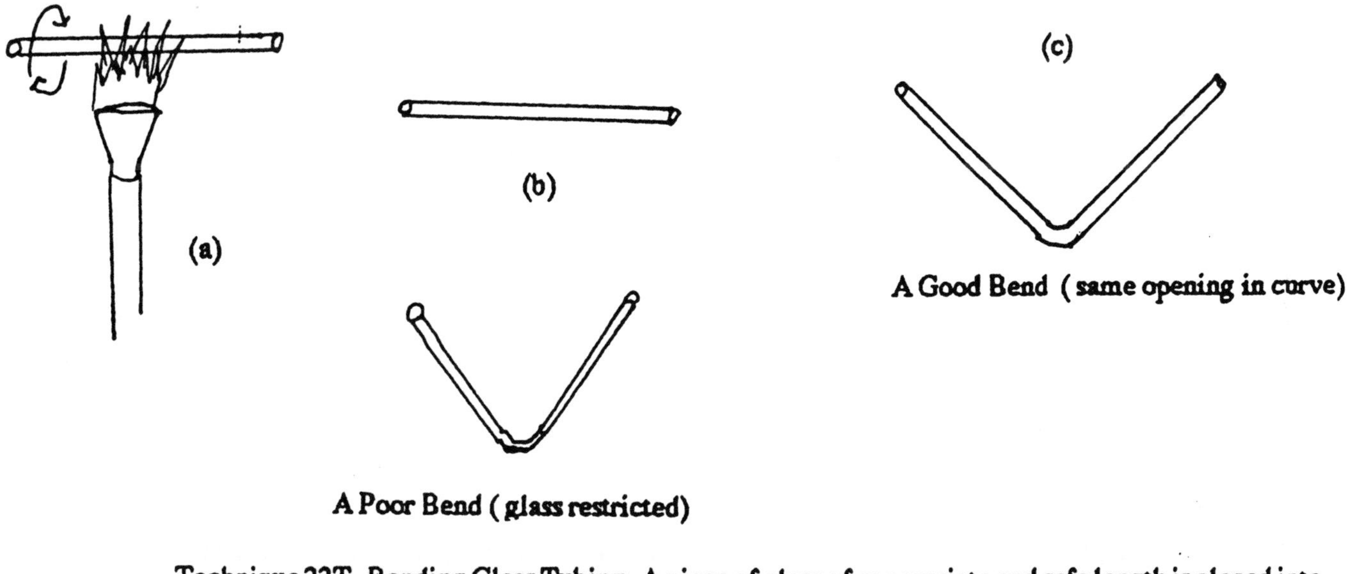

Technique 22T. Bending Glass Tubing. A piece of glass of appropriate and safe length is placed into a "hot" bunsen flame between the fingers of your hands with a wing top on the burner and rotated (a). Once the glass softens which is evidenced by the development of the yellow flame characteristic of sodium appears. The glass is quickly removed for a few seconds, (b) followed by bending into the desired shape (angle) (c).

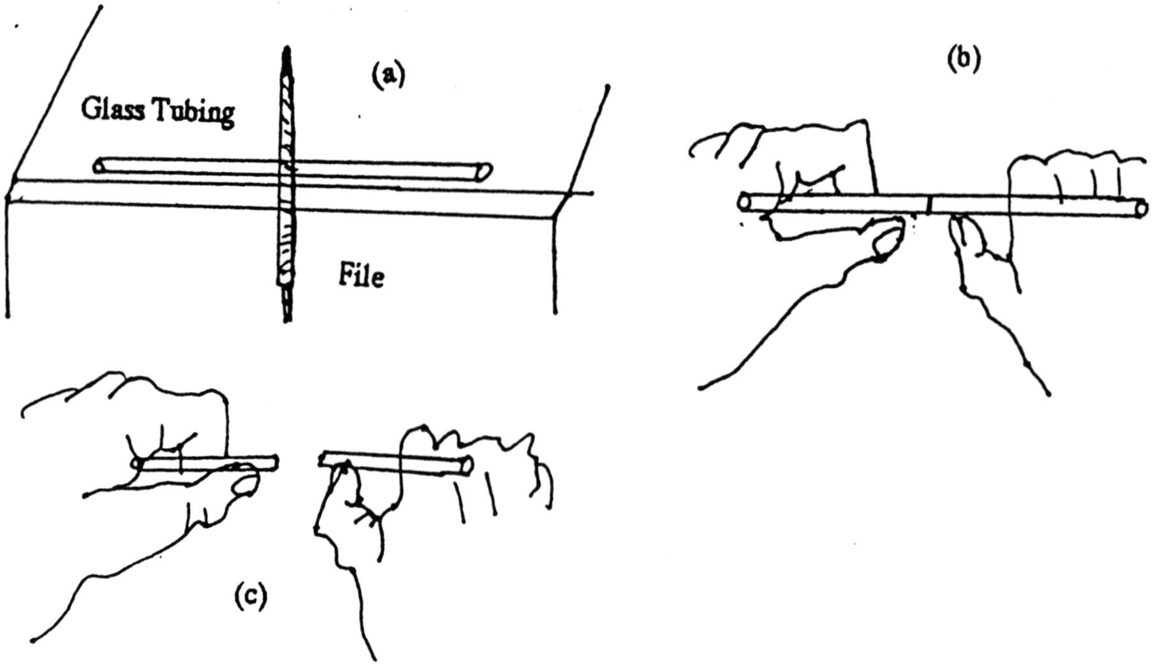

Technique 23T. Cutting Glass Tubing. Using a triangle file, make a single scratch mark onto a piece of glass tubing while holding it at the edge of the lab bench top (a). Pick up the glass and rotate the scratch away from you holding your thumbs on opposite sides of the scratch (b). Wet the crack and apply a slight outward pressure (c). The glass should snap apart nicely. The ends of the tubing should now be fire polished.

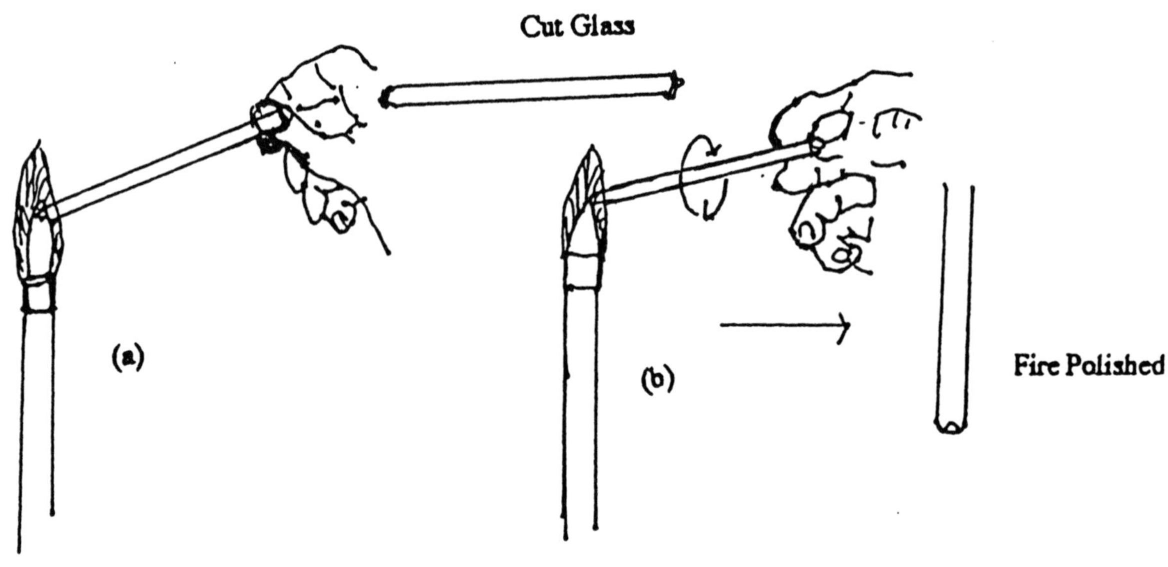

Technique 24T. Fire Polishing. Using a "hot" bunsen flame, the cut glass end is inserted into the flame (a). After considerable warming, the cut end is rotated back and forth until the edges are rounded and smooth (b).

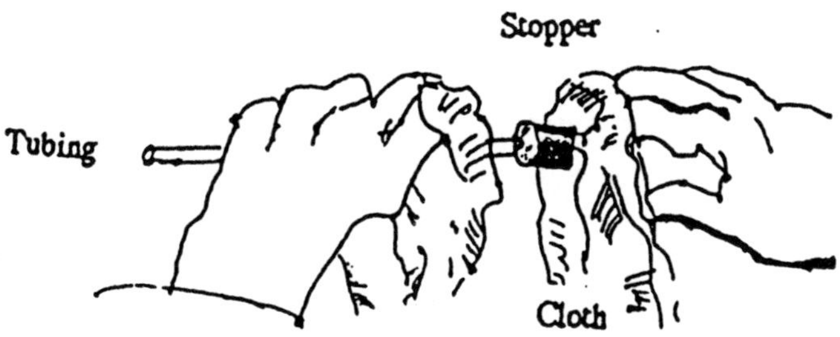

Technique 25T. Inserting a Glass Rod or funnel into a Rubber Stopper. Moisten the rubber stopper with water or mineral oil if appropriate and insert with a twisting motion. Excessive force should not be used when inserting the tubing. Always use a cloth or other protection so as to minimize cuts. Never insert without protection for the hands.

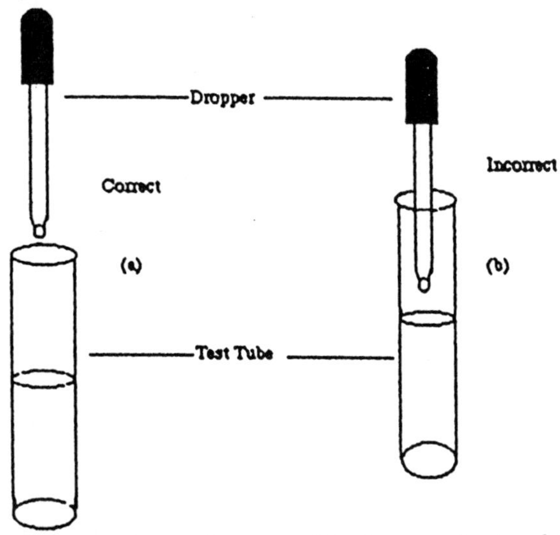

Technique 26T. Using a Medicine Dropper. The dropper is placed just above the mouth of the test tube (a). The drop or drops of liquid should be dropped in the middle of the test tube unless otherwise directed. The dropper is never inserted into the test tube (b).

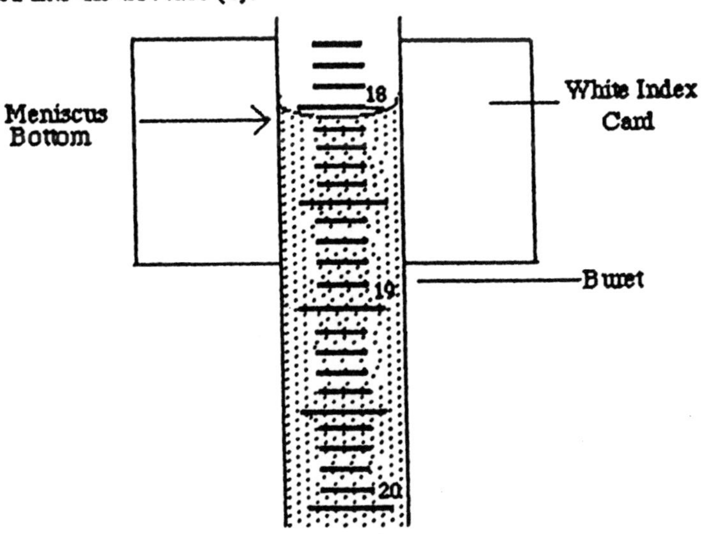

Technique 27T. A Buret is cleaned by adding a soap solution, followed by brushing with a buret brush. The buret is placed over a sink and rinsed with tap water(x3), followed by distilled water(x2). Finally the buret is rinsed with the desired reagent. To read the buret, place a piece of white index card or paper behind the buret and read it from the bottom of the meniscus for most liquids looking through the center of the buret. The buret can be read to the 0.1mL place and estimated to 0.01mL.

LABORATORY CHECK-IN

Each student will be assigned a laboratory **drawer** in the laboratory. You will be responsible for the drawer contents. A check-in sheet will be given to you. The instructor will identify each of the drawer items. Any item found missing will be replaced after all the items have been specified. Your signature is required to check-in the laboratory. The items in the **cabinet** will also be identified, however they are for the use of all the lab sections of the General Chemistry courses.
Several of the important Techniques and Other procedures will be presented and discussed thoroughly during the laboratory Check-In period. Included among these are:

1. General Laboratory Rules and Regulations
2. Using a Gas Lighter
3. Lighting a Bunsen Burner
4. Flame Adjustment (Cool and Hot Flames)
5. Cutting Glass Tubing
6. Bending Glass
7. Reading a Buret
8. Using a Medicine Dropper
9. Reading the Volume of a liquid
10. Weighing on the Electronic Balance
11. Fire Polishing
12. Safety Procedures and Role of Stockroom Personnel

EXPERIMENT 1

MEASUREMENTS

OBJECTIVES

a) To accurately measure mass, length, and temperature using SI units
b) To distinguish between fundamental and derived measurable quantities
c) To determine the following derived quantities, volume, density, and concentration
d) To understand and determine the accuracy, and precision of replicate measurements

SUPPLIES AND EQUIPMENT

Top-loading electronic balance (to 400 g max, 0.01 g smallest unit of mass), graduated cylinders 10 and 50 mL, Several 1.00 L graduated cylinders (for a class of 20), KNO_3 and $KClO_3$, 5.0 mL plastic syringes, a porcelain pestle, 10 wooden rulers graduated into cm, wooden meter sticks, vernier calipers, Quikcrete Concrete Mix, Aunt Jemima Syrup, Prestone Antifreeze, Dawn Dishwashing Detergent, Revlon Flex Balsam Shampoo and a Thermometer, -10 to 100°C, Graph Paper

Safety Concerns- Freshly mixed cement or concrete may cause skin injury. Avoid contact with skin. If contact occurs, wash skin promptly with water.

WHAT YOU SHOULD KNOW BEFORE YOU BEGIN

A measurement is a special kind of observation that scientists frequently make and it always compares some quantity (the seven fundamental measurable quantities are listed in Table 1) to a standard quantity. Our lab, and many others, uses a set of standards belonging to the International System of Units (SI Units). These are listed in Table 1.1 We can use the base units in Table 1 or attach prefixes to these units to represent fractions or multiples of the base units. You'll find some common prefixes in Table 1.2. Perhaps you were surprised that there were so few fundamental measurable quantities. However by combining several fundamental quantities we obtain derived measurable quantities and there are literally hundreds of these.

Table 1.1 SI Fundamental Measurable Quantities and Units

Physical Property	Name of Unit	Symbol
Length	meter	m
Mass	kilogram	kg
Time	second	s
Electric current	ampere	A
Temperature	kelvin	K
Luminous intensity	candela	cd
Amount of substance	mole	mol

Table 1.2. Common Prefixes used in SI Measurement System

Prefix	Symbol	Factor
mega	M	10^6
giga	G	10^9
kilo	k	10^3
deci	d	10^{-1}
centi	c	10^{-2}
milli	m	10^{-3}
micro	μ	10^{-6}
nano	n	10^{-9}

Table 3 lists some important derived fundamental quantities and some representative units for their measurement. For example, to measure speed, a derived quantity, you must measure both length and time, two fundamental quantities.

Table 3. Some Derived Measurable Quantities

Quantity	Units	Fundamental Quantities Measured
Area	sq. meter (m^2)	length, length
Volume	cubic meter (m^3)	length, length, length
Speed	meter/second (m/s)	length, time
Density	kilogram/cubic meter(kg/m^3)	mass, volume

In order to increase the reliability of a measurement we frequently make repeated measurements of the same quantity, we call these replicate measurements. We try to do the same manipulations the same way each time we make a measurement. When we've made our replicate measurements which measurement is best? We don't know, however we often use the average of the measurements as the best value. When we discuss how well a measurement was made we use the term **accuracy.** Accuracy refers to the closeness of a measurement to the true value. Well, how do we know what the true value of a measurement is (for example what is the true value of the width of your lab bench)? The true value is the average of a large number (> 20) of measurements made on a quantity when these measurements are made with care and with every effort to eliminate uncertainty. These true values are then published in textbooks. Your lab book has an Appendix containing "true values" for such quantities as Avogadro's number, the speed of light, and the mass of an electron.

We have several common ways of describing accuracy; absolute error and percent error.

Absolute error = | true value - your value |

Percent error = Absolute error (100)/ True Value

Of course an accurate measurement has a low (usually less than 5%) percent error. **Precision** refers to a set of replicate measurements and is a description of the closeness of repeated measurements to each other. Precision and accuracy are not related. It's possible to be precise but inaccurate or imprecise and accurate. In today's experiment you should strive to be both accurate and precise. One way to measure precision is to calculate a quantity called the standard deviation which is described in your Appendix. A small standard deviation means good precision.

Now, what goes into making an accurate measurement? Basically three things; the measuring instrument and its quality, the measurer (that's you) and the procedure used to make the particular measurement. In today's experiment your instructor will show you all the measuring instruments you'll need and how to use them correctly so please pay attention.

Several of the derived quantities need further discussion. **Volume** is a measure of the space an object occupies in three dimensions and it has the units of length x length x length or length^3. For regularly-shaped objects there are well-defined relationships between the volume of the object and some distance measured on the object.
For example;

Regularly-shaped Solid	Formula for Volume
Rectangular	$V = l \times w \times h$
Sphere	$V = 4/3\pi r^3$
Cylinder	$V = \pi r^2 h$

where r= radius, h= height, w= width and l=length. For irregular-shaped objects no easy formulas are available, however often the volume of these objects can be measured by water displacement. Submerging an object in water will cause the level of water in the container to rise. The increase in the volume of the water containing the object will equal the volume of the object.

It's often useful to know the mass of a given volume of a sample of matter. With this knowledge you can get any desired mass of a substance by measuring out a given volume. The relationship between mass and volume for the liquid metal, mercury, appears below:

Mass of Mercury (g)	Volume of Mercury(mL)	Mass/Volume
27.2	2.00	13.6 g/mL
122.4	9.00	13.6
150	9.20	13.6
272	20.0	13.6

Notice the volume increases as the mass increases. Dividing the mass by the volume gives us the mass of a unit volume (1 mL) and this is the derived quantity we call **density.** When we comment about how "heavy" an object is we're usually referring to the object's density. When we have two or more liquids of different densities the less dense liquid will always "float" on top of the more dense liquid. This is why huge oil spills in the ocean are such a problem, the oil is less dense than the water and stays on top of the water causing problems for birds and other mammals.

When we mix two or more substance to form a homogeneous mixture we have a **solution**. But how do we describe how much of each substance is present in the solution? We do this by referring to the **concentration** of the solution, a term which tells us the amount of the **solute** (the minor component of the mixture) and the amount of the **solvent** (the major component of the mixture). For example, if we dissolve 5.0 g of tea in 95 g of water to make 100 g of solution we can say the solution is 5% tea by mass. We have expressed this concentration as a **mass percent** (commonly called weight percent) a common way of reporting concentrations. The maximum weight of a solute that a solvent can hold at a given temperature is called the **solubility** of the solute in that solvent. The solution is said to be saturated with the solute. Solubility varies with temperature, generally the higher the temperature, the more solute will dissolve in a given mass of solvent.

EXPERIMENTAL PROCEDURE

Today's procedure is purposely vague to allow you and your partner (yes, you will need a partner to complete today's experiment) to use your own judgement in completing the assigned task. If you need help, ask the instructor for assistance or simply watch what some of your classmates are doing. That will get you started.

A. A. WOODEN BLOCK MEASUREMENTS (X3)

1. Obtain a wooden block and carefully weigh it to the nearest 0.01g.
2. Measure the width, length, and height of it using a centimeter ruler ($\pm$ 0.1cm). Be sure to keep the same sides constant as you take the measutrements.
3. Measure the width, length, and height of it using a vernier caliper ($\pm$ 0.01cm).
4. Calculate the volume in cm^3 using the V= L x W x H formula.
5. Calculate the Density , Density (ave) and the Standard Deviation (Appendix D).

B. SALT SOLUBILITY AT VARIOUS TEMPERATURES

4. Set up the apparatus in Figure 1.1 using your large test tube. A thermometer is inserted through one hole in the stopper and a copper wire stirrer is inserted through the second hole. Thermometers, stirrers and stoppers can be obtained from the storeroom. The wire loop should be around the thermometer. Weigh the assembly (test tube, thermometer, stopper and stirrer) to the nearest 0.01g then add the amount of **one** salt indicated in the table below;

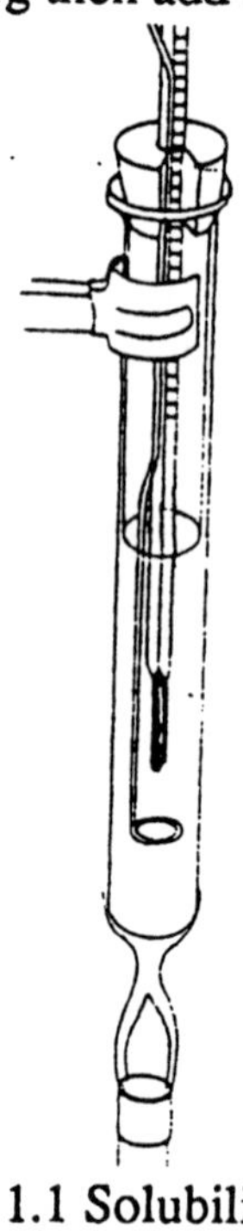

Figure 1.1 Solubility-Salt Apparatus

5. Accurately (0.01g) weigh the assembly and the added salt and determine the mass of the salt by subtraction.

6. Be sure to read the label so that you know which of the two salts you are using. Write down the formula. Using your plastic syringe, add the indicated volume of water then heat the test tube and contents while stirring. Make sure the tip of the thermometer is near the bottom of the test tube and is immersed in the water. Heat until all the solid dissolves then remove the heat but continue to stir. Carefully watch the solution for the first signs of the appearance of small crystals of the salt. When solid first appears, record the temperature on your report sheet.

Salt	$KClO_3$	KNO_3
Weight of salt	4.5-5.0g	5.5-5.7g
1st vol. water	10 mL	4 mL
2nd vol. water	2 mL	2 mL
3rd vol. water	8 mL	2 mL
4 th vol. water	5 mL	4 mL
5th vol. water	5 mL	4 mL

7. Cool the assembly to room temperature, accurately measure and add a second portion of water, and repeat the heating and cooling cycle again noting the temperature at which crystals just begin to form. Repeat for a total of five measurements.

8. Record all data on your report sheet and calculate the concentration (g salt/ 100 g solvent) of the saturated solutions at each temperature.

9. Finally, on the attached graph paper, plot the temperature on the horizontal axis and the concentration (g salt/ 100 g water) on the vertical axis. Remember to properly label your graph.

C. CONCRETE MEASUREMENTS

10. Obtain and weigh (0.01g) a styrofoam coffee cup.

11. Accurately weigh about 80 g (± 0.01) of Quikcrete Concrete Mix into the coffee cup and add enough water to make a thick paste. **Mix this stuff well.**

12. Accurately weigh the cup and contents again. Now let the container sit until the concrete solidifies (two to three days).

13. Once the concrete has hardened, remove the coffee cup and, using a ruler, measure the height and diameter of the concrete cylinder.

14. Next weigh the concrete cylinder (0.01g).

15. Calculate the volume and density of the concrete. $V = 1/3\pi h(R_1{}^2 + R_1R_2 + R_2{}^2)$
To check your volume measurement, determine the volume of the concrete a second time by water displacement.

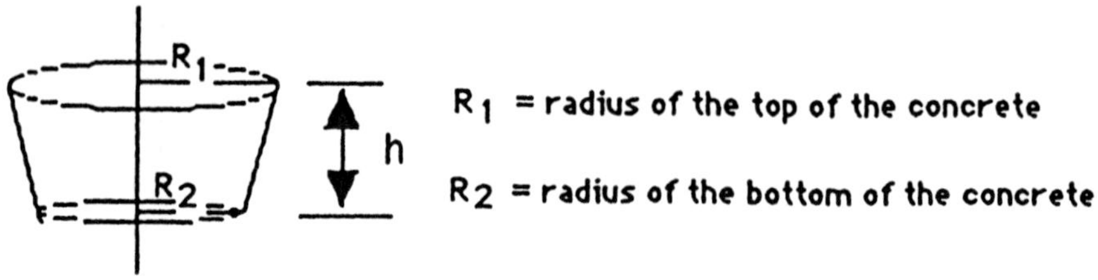

16. After removing the concrete from the water, take a paper towel and wipe off the surface water. Weigh the concrete cylinder again. By the way, what do you know about concrete?

SOME CONCRETE FACTS

American yearly produces two tons of concrete for every person in the U.S., mostly because concrete is strong, cheap, versatile and easy to make. Concrete is a mixture of sand, gravel and crushed stone held together by water and cement. Cement is a fine powder of limestone ($CaCO_3$) and clay (an aluminosilicate). When concrete hardens, a complex series of chemical reactions occur in which the cement and water react to form calcium and silicate hydrates which grow forming a set of interlocking crystals which hold together the entire mass. The hydrate crystals continue to grow increasing the concrete's strength over a period of years. How durable is concrete? Depending upon the chemical environment surrounding the concrete it can survive for millions of years or less than one year. Concrete has pores and water can enter these pores through cracks in the concrete. When this happens and the water freezes and expands, it can burst the concrete apart. Also certain bacteria convert waste sulfides to sulfuric acid, an extremely corrosive substance, which literally dissolves concrete.

D. DENSITY OF DIFFERENT SOLUTIONS

17. Observe the 250 mL graduated cylinder your instructor has set up. This cylinder contains the following solutions, Aunt Jemima Syrup, Prestone Antifreeze, Dawn Dishwashing detergent, Revlon Flex Balsam Shampoo and water. Which solution has the greatest density? Which has the smallest density?

18. Add a mL of water, a small cork and a piece of plastic cut from an empty plastic milk container to the graduated cylinder. Describe what happens.

EXPERIMENT 1 MEASUREMENTS

DATA AND LAB REPORT SHEET

NAME____________________________ASSIGNED #_____ DATE__________SCORE______

PARTNER__________________________ASSIGNED #_____

A. WOODEN BLOCK

	Trial 1	Trial 2	Trial 3
Mass	__________g	__________g	__________g
Width (Ruler)	__________cm	__________cm	__________cm
Width (Caliper)	__________cm	__________cm	__________cm
Length (Ruler)	__________cm	__________cm	__________cm
Length (Caliper)	__________cm	__________cm	__________cm
Height (Ruler)	__________cm	__________cm	__________cm
Height (Caliper)	__________cm	__________cm	__________cm
Volume(cm^3)	__________cm^3	__________cm^3	__________cm^3(Ruler)
Volume(cm^3)	__________cm^3	__________cm^3	__________cm^3(Caliper)
Density	__________g/cm^3	__________g/cm^3	__________g/cm^3(From Ruler calc.)
Density	__________g/cm^3	__________g/cm^3	__________g/cm^3(From Caliper calc.)

Density Ave. ________g/cm^3 (From Ruler) Standard Deviation ________

Density Ave. ________g/cm^3 (From Caliper calc.) Standard Deviation ________

B. SALT SOLUBILITY AT VARIOUS TEMPERATURES

a) Wt. of Assembly + Sample of Salt ____________ b) Wt. of Assembly_____________

c) Wt. of Salt_______________

Total Volume of Water (0.1ml)	Saturation Temp.	Salt Conc. (g/100g Solvent)
__________________	______________	__________________
__________________	______________	__________________
__________________	______________	__________________
__________________	______________	__________________
__________________	______________	__________________

Sample Calculation of Salt Solubility (g/100g of Solvent) For One Trial

C. CONCRETE MEASUREMENTS

a) Mass of Coffee Cup ____________________ Mass of Cup & Mixture ____________________ Mass of Hardened Concrete ____________________

Mass of Water Loss________________________

b) Measurement of Concrete Cylinder

Height ______________ Diameter ___________ Calc. Volume_________

Volume by Water Displacement _____________ Mass After Water Displacement __________

Density (Mass/Calc. Volume)__________________

Density (Mass/Measured Volume)_______________

Best Density Value_____________ Explain.

D. DENSITY OF DIFFERENT SOLUTIONS

a) Which Solution Has The Greatest Density? ________________

Which Solution Has The Smallest Density?________________

b) Describe What Happens To The:

Tygon Tubing -

Small Cork-

Piece of Plastic-

POST-LAB QUESTIONS

1. Which measurements of the length of the wooden block are more accurate? more precise?

2. What is your most accurate value for the density of the concrete? Explain your answer.

3. How would you change the experimental procedure if you wished to measure the solubility of your salt at $0^{o}C$?

4. Did you find a difference between the volumes of the concrete cylinder determined by use of the formula and by water displacement? How do you explain this difference?

5. Now that you've completed the experiment list several things you think we should do to improve it.

Pre-Laboratory Assignment Experiment 1
Measurements Name________________________________#_____ Score_______

1. Three students repeatedly measured the mass of a block of steel.
   (Student A, 35.3, 35.4, 35.5 g, Average = 34.9, True value= 34.6)
   (Student B, 36.5, 35.9, 32.1 g, Average = 34.9, True value= 34.7)
   (Student C, 34.4, 34.5, 34.4 g, Average value = 34.4, True value= 34.4).

   Student A data is of ________precision and _______ accuracy. (Use low and high).
   Explain:

   Student C data is of ________precision and _______ accuracy. (Use low and high).
   Explain:

   Student B data is of ________precision and _______ accuracy. (Use low and high).
   Explain:

2. Using SI units for the fundamental quantities in what units would the following quantities be expressed?
   a) mass b) volume c) temperature d) density

3. What problems would you encounter if you attempted to determine the volume of a rectangular-shaped sponge by water displacement? Would your measured volume be larger or smaller than the true volume of the sponge? Explain.

4. How many separate measurements are made of the width of the wooden block?

5. 1.50g of $NaHCO_3$ is dissolved in 17.1g of water. Express the solubility of $NaHCO_3$ in grams per 100g of solvent and calculate the % composition of $NaHCO_3$ (solute) and water (solvent). Density of water = 1.00g/mL @25°C.

10 DIVISIONS PER CENTIMETER

EXPERIMENT 2

IDENTIFICATION OF SUBSTANCES:

OBJECTIVES:

a) To become familiar with techniques used in the determination of the density, melting or boiling points, and solubility of substances
b) To use several physical properties to identify an unknown substance

Note To the Instructor: Students should be assigned a liquid or a solid unknown. For any solid, the melting point can be determined with the boiling point apparatus using water or can be performed with the Electrothermal Melting Point Apparatus. The boiling point of the liquid should be determined using water. Therefore do not assign a liquid with a boiling point greater than that of water. The solid cannot be soluble in both solvents in the experiment for density determination unless alternate is used.

SUPPLIES AND EQUIPMENT

An unknown solid or liquid, boiling point (Technique 18T) or melting point apparatus (Technique 19T) as dictated by unknown designation, mineral oil, capillary tubes, rubber bands, 150 ml beaker, wire gauze, ring stand, test tubes, 50 ml graduated cylinder, rubber stoppers, thermometer, three prong clamp, ethanol , Electrothermal Melting Apparatus for Melting point of solid unknowns.

WHAT YOU SHOULD KNOW BEFORE YOU BEGIN

Matter, all components of which have the same properties under the same physical conditions, and which has the same chemical composition is regarded as a substance. A substance can be an element, or a compound. It can be identified by **chemical properties** observed when the substance undergoes chemical changes or by **physical properties** which are often performed. The physical properties investigated in this experiment include, solubility, density, and melting or boiling point.

The **solubility** of a substance is referred to as it's ability to dissolve in a pure solvent at a specified temperature. In relative terms, a solute is characterized as soluble (s, completely dissolved), slightly soluble (sl s, partially dissolved) , and insoluble (undissolved) . Quantitatively, solubility can be measured. In the "Handbook of Chemistry and Physics." Solubilities are listed for substances in several solvents. For inorganic substances, it is listed as grams of solute/100cc of cold and hot water at a specified temperature. For organic substances, it is listed as relative solubility (v, very soluble; s, soluble; i, insoluble; slightly soluble). The solubility of liquids in liquids is often referred to as the miscibility of substances (ability to mix) . Soluble substances are said to be miscible, while insoluble substances are said to be immiscible. In this experiment, the relative solubility will be referred to as s, sl s, and i.

The **density** refers to the compactness of matter (mass per occupied space), and is expressed as a mass of substance per unit volume ($D = M/V$). Pure substances can be partially identified by their densities. The "Handbook of Chemistry and Physics" list the densities of substances.

The **melting point** is defined as the temperature at a fixed pressure at which a substance's solid and liquid states are at equilibrium. A solid will be converted into a liquid or it may decompose. For pure substances the melting point occurs at one specific temperature, while mixtures of solids melt over a range of temperatures, which can be as much as 4-5°C.

The **boiling point** is the temperature at which the vapor pressure of the surroundings is exactly equal to that of the liquid. The vapor can then escape from the surface of the liquid. Rapid bubbling will be observed.

EXPERIMENTAL PROCEDURE:

You will be assigned an unknown solid or liquid substance. You will measure the solubility, density and melting point or boiling point of the unknown. Once the measurements and determinations have been made, a comparison of your results to the physical properties in Table 2.1. of substances should be done to ascertain the identity of the unknown substance. Obtain approximately 16 g of solid unknown, or 15 ml of liquid unknown and record the unknown number or letter on the report sheet.

Solid:

1. **Solubility:** Set up two test tubes in a test tube rack , and pour 20 drops of water in one, 20 drops of ethanol in the other. To each tube add several crystals about the size of a match head of the unknown. Stopper each test tube and shake vigorously for 1 minute, and allow the test tube contents to settle. Carefully observe whether solid has dissolved, and record observations for each solvent on the Data-Report Sheet, **(1)**. Record as "i", insoluble, "s", soluble, or "sl s" , for slightly soluble.

2. **Density:** Weigh to the nearest 0.1g approximately 5.0grams of unknown solid, **(2)**. Pour approximately 15 ml of solvent that the solid was found insoluble in, into your 50mL graduated cylinder , read and record the volume of the liquid to the nearest 0.1ml, **(3)**. Add the solid to the cylinder carefully not spilling any of the solid. Stir with a glass rod to remove air bubbles. Read and record the new volume of the cylinder plus unknown, **(4)**. The difference in volumes represents the volume occupied by the solid, **(5)**. Repeat for trial two by adding an additional 5.0mL of solvent and 1.5g of solid. Repeat for trial three by adding 5ml more of solvent and 1.50g of solvent. Calculate the density of the solid unknown, (D= M/V), **(6)**; average density, **(7)**; and the standard deviation, **(8)**.

   Alternate Density Procedure: Pour a small sample of the solid unknown into a pellet press, close and apply pressure to form a solid disc. Remove the disc and determine the dimensions using a micrometer. Weigh and record the mass of the disc, **(A1)** diameter,**(A2)** and the thickness of the disc, **(A3)**. Using the formula for the volume of a circular cylinder, $V= \pi r^2h$ calculate the volume of the disc, **(A4)**. Convert the volume of the pellet to units of cm^3, **(A5)**. Calculate the density of the solid unknown, (D= M/V), **(A6)**; average density, **(A7)**; and the standard deviation, **(A8)**.

3. **Melting Point:** Obtain a capillary tube with one end sealed. Place a few crystals of unknown into a mortar and pestle and grind to a powder. Touch the open end of the capillary tube onto the powder and obtain a very small amount of solid into the end of the tip. Only enough for you to see. Remove the capillary and obtain a glass tube approximately two feet in length and drop the capillary down the inside of the glass tubing with the closed end first. Repeat three times. Place the capillary in the boiling point apparatus or in the Electrothermal Melting Point Apparatus (Follow instructions on use). If using the boiling point apparatus, pour water into the apparatus as designated and mount the capillary along side the thermometer and clamp in place in the water bath. Heat with a low flame. Carefully observe the solid and record the melting point range. Note and record exactly when the crystals first begin to melt and the temperature at which they all have melted, **(9)**. Often the purity of substances is ascertained by the shortness of the temperature range. Repeat by allowing the water to cool or add additional water.

   Safety Precautions: 1. Handle the apparatus only when cool.
   2. Wear safety glasses at all times.

Liquid:

1. **Solubility**: Set up two test tubes in a test tube rack , and pour 20 drops of water in one, 20 drops of ethanol in the other. To each tube add 20 drops of the unknown. Stopper each test tube and shake vigorously for 1 minute, and allow the test tube contents to settle. Carefully observe whether the liquid has dissolved, partially dissolved, or did not dissolve and record observations for each solvent on the Data-Report Sheet. Record as "i", insoluble, "s", soluble, or "sl s" , for slightly soluble, (**1L**).

2. **Density:** Weigh to the nearest 0.01g a 10 ml graduated cylinder and record it on the report sheet, (**2L**). Pour approximately 6.5 ml of unknown liquid into the cylinder, read the bottom of the meniscus and record the volume to the nearest 0.1ml, (**3L**). Re-weigh the cylinder and liquid and record the weight,(**4L**) . Determine the weight of the liquid (**5L**). Calculate the density of the unknown liquid,(**6L**); average density, (**7L**); and the standard deviation, (**8L**). To repeat for a second and third trial, simply adding more unknown liquid. Do not add more than 10ml.

3. **Boiling Point:** Set up a Boiling point apparatus as shown on Page 15. Attach a small centrifuge test tube to a thermometer with a rubber band. Add 15 drops of unknown liquid into the test tube. Place test tube and unknown into the apparatus containing 75ml of water. Place an inverted capillary tube into the test tube (closed end out). Heat with a low flame until the liquid in the test tube boils gently, and bubbles appear at the mouth of the capillary tube. Once it boils, turn off the bunsen burner and observe exactly when the bubbles just stop , before any liquid enters the capillary tube. Read and record the temperature as the boiling point of the liquid to the nearest 1.0^oC,(**9L**). Repeat by allowing the liquid to cool down. Determine the boiling point average, (**10L**). It may be necessary to add a few drops of unknown.

Precautions:
1. Do not allow any unknown to drop into the hot water.
2. Handle the apparatus only when cool.
3. Wear safety glasses at all times during the experiment.

EXPERIMENT 2 IDENTIFICATION OF SUBSTANCES

DATA AND LAB REPORT SHEET

NAME______________________ ASSIGNED #________ DATE___________

Solid Unknown:
Unknown Number or Letter________ *

		Water	Ethanol
Solubility of unknown	1.	______	______

		Trial 1	Trial 2	Trial 3	
Density:	2. Mass of solid	________	________	________	g
	3. Volume of liquid	________	________	________	ml
	4. Volume of liquid + solid	________	________	________	ml
	5. Volume of Solid	________ (4-3)	________	________	ml
	6. Density of Unknown	________	________	________	g/ml
	7. Density Average		___________ g/ml		
	8. Standard Deviation		___________		

Melting Point:	**Begins**	**Ends**	
9. Melting Point Range	____________	____________	°C

Show All Calculations On Lineless Paper: Density, D Ave, Standard Deviation

Identity of the unknown: ________
Your unknown is one of the substances in Table 1. Using the experimental data, determine the identity of the unknown by comparison. Select the best fit using all the properties examined.

Rationale:

Alternate Density of the Solid

		Trial 1	Trial 2	Trial 3	
11.	Mass (Pellet)	_______	_______	_______	g
12.	Diameter(Pellet)	_______	_______	_______	m
13.	Thickness, h	_______	_______	_______	m
14.	Volume (Pellet)	_______	_______	_______	m^3
15.	Volume	_______	_______	_______	cm^3
16.	Density of Unknown	_______	_______	_______	g/ml
17.	Density Average		_______ g/ml		
18.	Standard Deviation		_______		

Liquid Unknown:
Unknown Number or Letter________ *

		Water	Ethanol
Solubility of unknown	1L.	_________	_______

			Trial 1	Trial 2	Trial 3	
Density:	2L	Mass of cylinder	_________	________	________	g
	3L.	Volume of liquid	_________	________	________	ml
	4L.	Mass of cylinder + liquid	_________	________	________	g
	5L.	Mass of liquid	_________	________	________ (3-2)	g
	6L.	Density of Unknown	_________	________	________	(g/ml)
	7L.	Density Average		___________(g/ml)		
	8L.	Standard Deviation		___________		

Boiling Point:

9L. _________ ________

10L. Boiling Point Average __________

Show All Calculations On Lineless Paper: Density, D Ave, Standard Deviation

Identity of the unknown: ________
Your unknown is one of the substances in Table 1. Using the experimental data, determine the identity of the unknown by comparison. Select the best fit using all the properties examined.
Rationale:

Table 2.1: PHYSICAL PROPERTIES OF SELECTED SUBSTANCES

Name	Boiling Point	Melting Point	Density g/ml	Solubility Water	Solubility Ethanol
	°C	°C			
Acetone	56	-95	0.79	s	s
2-butanone	80	-87	0.81	s	s
Cadmium nitrate · $4H_2O$	132	59	2.46	s	s
Cyclohexane	81	6.5	0.78	i	s
Cyclohexene	83	-104	0.81	i	s
p-dibromobenzene	219	87.3	1.83	i	s
p-dichlorobenzene	174	53	1.46	i	s
Diphenyl	255	70	0.99	i	s
Diphenylamine	302	54	1.16	i	s
Diphenylmethane	27	265	1.00	i	s
Ethanol	79	-117	0.79	s	s
Ethylacetate	77	-84	0.90	sl s	s
Heptane	98	-91	0.68	i	s
n-Hexane	69	-94	0.66	i	s
Isopropyl alcohol	83	-89.5	0.79	s	s
Lauric acid	161	48	0.87	i	s
Magnesium nitrate· $6H_2O$	330c	89	1.63	s	s
Methanol	65	-98	0.79	s	s
Naphthalene	218	80.2	1.16	i	s
n-Propanol	97	-127	0.81	s	s
Propionaldehyde	49	-81	0.81	s	s
Sodium acetate ·$3H_2O$	123	58	1.45	s	sl s
Zinc chloride	732	283	2.91	s	s
Zinc nitrate · $6H_2O$	36	105d	2.06	s	s

c = boils upon decomposition, d = decomposes, s = soluble,
sl s = slightly soluble, i = insoluble

Pre-Laboratory Assignment Experiment 2
Identification of Substances Name______________________ #____ Score______

1. What is meant by the term"Solubility?"

2. What is the density of an unknown if 6.5g of it occupies a volume of 2.5mL?

3. Could you determine the density of zinc chloride using water as a solvent? Explain

4. What is meant by the term "physical property?"

5. An unknown was found to have a density of 0.65g/ml and a boiling point of 68.5^{o}C. It was also soluble in ethanol and insoluble in water. What is the identity of the unknown?

EXPERIMENT 3

FORMULA OF A COMPOUND

OBJECTIVES

a) To determine the empirical formula of a compound
b) To investigate the formation of a compound
c) To perform percentage composition calculations

SUPPLIES AND EQUIPMENT

Mg ribbon, Crucible, Triangle pipestem, Bunsen burner, Tongs, Balances, Dessicator

WHAT YOU SHOULD KNOW BEFORE YOU BEGIN

A chemical compound is the direct result of two or more elements combining together. Chemical formulas are used to represent compounds. Before we can learn how chemical formulas are written, we must first acquaint ourselves with the symbols used to denote the elements from which these compounds are formed. Many symbols are derived from the first syllable of the elements' name. We can combine the symbols to denote the formulas of compounds. According to the Law of Constant Composition (Law of Definite Proportions), a given chemical compound has the same relative ratio by weight of the elements that constitute the particular compound.

A mole of any element is defined as the amount of that element that contains the same number of atoms, ions, or molecules as exactly 12g of ^{12}C. It has been determined experimentally that the number of atoms, ions, or molecules in this quantity of ^{12}C is 6.022×10^{23}. This number is given a special name, **Avogadro's number.** This number of atoms, ions, or molecules or formula units weighs one gram atomic weight, one gram ionic weight, one gram molecular weight or one gram formula weight.

After a chemist has determined by qualitative analysis , which elements are present in the compound, there is a need to know the empirical formula of a compound, i.e. relative number of moles of each element in the compound. The **empirical formula** is the simplest whole number ratio of the atoms of elements to each other in a mole or formula unit of a compound. An empirical formula can be calculated from the percentage composition which indicates the relative weights of the constituent elements. **Empirical weights** are determined from empirical formulas by taking the sum of (the ratio of the atoms of elements times the atomic weight of the elements). The formula NaCl implies that there are 23 grams of sodium for each 35.45g of chlorine. In NaCl, the percent of sodium and chlorine by weight are :

$$\frac{\text{Mass of sodium}}{\text{Total mass of salt}} \times 100 = \frac{23.0}{(23.0 + 35.45)} \times 100 = 39.3\% \text{ Na}$$

$$\frac{\text{Mass of chlorine}}{\text{Total mass of salt}} \times 100 = \frac{35.45}{(23.0 + 35.45)} \times 100 = 60.7\% \text{ Cl}$$

We then determine the chemical formula of a compound using the simplest whole-number ratio. For example, an analysis of 70.0 grams of table salt contains 27.5 grams of sodium and 42.5 grams of chlorine. The moles of each element is calculated using their respective atomic weights.

27.5g Na x ($\frac{\text{1mole Na}}{\text{(23.0 g)}}$) = 1.20 moles Na

42.5g Cl x ($\frac{\text{1 mole Cl}}{\text{(35.45)}}$) = 1.20 moles Cl

The mole ratio for sodium chloride is $Na_{1.2}Cl_{1.2}$. However, elements combine in ratios of small whole numbers. Therefore this is determined by dividing each elements number of moles by the smaller value of all the elements present. The ratio of sodium to chlorine is reduced to the simplest whole-number ratio:

Sodium : 1.20moles / 1.20moles = 1 mole

Chlorine: 1.20 moles / 1.20 moles = 1 mole

Therefore, the formula of sodium chloride is NaCl. For each sodium , there is one chlorine. Although this is the chemical formula of sodium chloride, it must be pointed out that sodium chloride is an example of a type compound called an **ionic** compound which exists as a crystalline solid in a geometric array or lattice of ions.

Magnesium burns in air when heated. It combines with oxygen in the air forming magnesium oxide, and with nitrogen to form small amounts of magnesium nitride. If the reaction is carried out in the open atmosphere, the oxide will usually be lost to the air because it is quite low in mass. The unbalanced equation is shown below.

?Mg + ?O_2 ---> ?Mg_xO_y, the values of x and y will be determined experimentally. The x and y values are the moles of each element present in the chemical formula. The oxide of magnesium is a white solid that is ash-like in appearance.

The nitride balanced equation is : 3Mg + N_2 ---> Mg_3N_2 The nitride is coverted to the oxide by the addition of water followed by heating.

EXPERIMENTAL PROCEDURE

1. Clean and dry a crucible with lid.
2. Place it on the clay triangle as shown in Fig. 3.1 below.
3. Heat the crucible until it is dull red at the bottom.
4. Cool and record the mass of the crucible and lid (0.01g).
5. Place several pieces of Mg ribbon (0.16 - 0.20g), and place in the bottom of the crucible.

6. Re-weigh and record the mass of the crucible, lid and magnesium (0.01g).

7. Heat with a low flame with the lid on for approximately 15 minutes, periodically lifting the lid to allow air to enter the crucible. See Heating a Crucible Technique section .

8. Remove the lid with the tongs and heat the open crucible for 5 minutes.

9. Allow the crucible to cool for 10 minutes.

10. Add 10 drops of water and re-heat the crucible with lid slightly ajar.

11. Remove from the clay triangle and place on wire gauze on the desk top, and place a beaker over the crucible and cover.

12. Cool the crucible and lid, and re-weigh and record the mass (0.01g).

13. Repeat adding 10 drops of water, heating, cooling , and weighing until the masses agree within 0.02g

14. When removing the crucible after heating, place it in the dessicator.

Safety Precaution: Exercise care when handling the hot crucible.

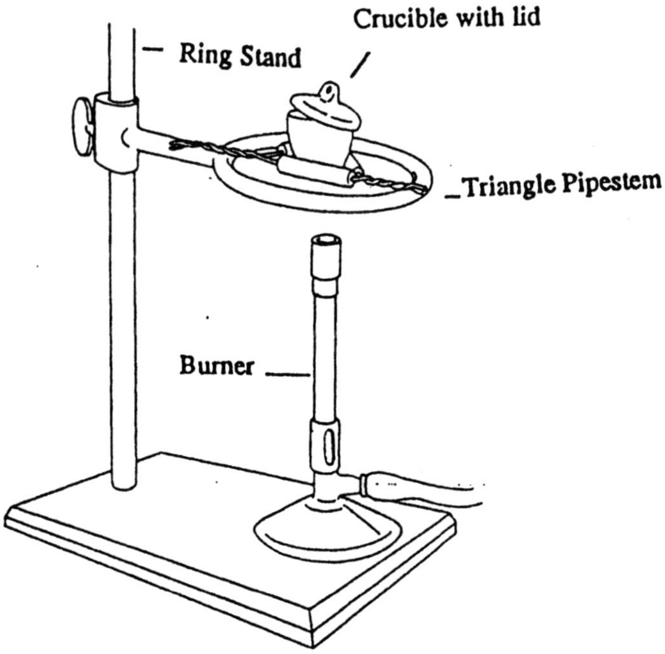

Figure 3.1 Heating Crucible Apparatus

EXPERIMENT 3 FORMULA OF A COMPOUND

DATA AND LAB REPORT SHEET

NAME______________________________ ASSIGNED #_____ DATE___________

Magnesium Oxide Synthesis	Trial 1	Trial 2
1. Mass of crucible and lid (g)	_______	_______
2. Mass of crucible , lid and Mg (g)	_______	_______
3. Mass of Mg (g)	_______	_______
4. Mass of crucible, lid, and compound (g)	_______	_______
a. 2nd weighing	_______	_______
b. 3rd weighing	_______	_______
c. 4th weighing	_______	_______
5. Mass of compound (g)	_______	_______
6. Mass of Oxygen in compound (g)	_______	_______
7. Number of moles of Mg	_______	_______
8. Number of moles of oxygen	_______	_______
9. Mole ratio of Mg to O	_______	_______
10. Formula of Magnesium oxide	_______	_______
11. Percent of Mg in compound	_______	_______
	___________ % Mg average	
12. Percent of O in compound	_______	_______
	___________ % O average	

SHOW ALL CALCULATIONS ON LINE LESS PAPER

QUESTIONS

1. What are possible sources of error in the experiment?

2. What is the balanced equation of the reaction of magnesium with oxygen?

3. What is the balanced equation of the reaction of magnesium with nitrogen?

4. What are the percent composition of each of the following substances?

$BaSO_4$ H_3PO_4

5. What are the empirical formulas and weights of each of the following?

$H_2C_2O_4$ $C_6H_{12}O_6$

Pre-Laboratory Assignment Experiment 3
Formula of a Compound Name________________________________#________Score ____

1. What is the mass in grams of 1 mole of carbon dioxide (CO_2), and silicon dioxide (SiO_2)?

2. What is the mass of 1.22×10^{23} molecules of carbon dioxide, and silicon dioxide?

3. Given the reaction $3Fe + 2\ O_2 \rightarrow Fe_3O_4$ which is ->[$FeO + Fe_2O_3$]

   a. How many moles of iron will react with 1 mole of O_2?

   b. How many moles of iron will react with 10 moles of O_2?

   c. How many moles of oxygen molecules (O_2) will react with 1 mole of Fe?

4. Determine the simplest formula of a compound that has the following percentage composition: 26.52 % Cr, 24.52 % S, and 48.96% oxygen. Use: Cr-52, S-32, and O-16 masses

EXPERIMENT 4

FORMULA OF A HYDRATE

OBJECTIVES

a) To understand the association of water as water of hydration in a compound
b) To write the formula of a compound with the water of hydration
c) To experimentally determine the number of moles of water per mole of unknown compound

SUPPLIES AND EQUIPMENT

Hydrate salts, Crucible and cover, Triangle Pipestem, Bunsen burner, Tongs, Wire Gauze

WHAT YOU SHOULD KNOW BEFORE YOU BEGIN

Many substances chemically combine with water to form dry, crystalline compounds. These compounds, containing a salt and water combined in definite proportions are known as hydrates; the water associated with them is called **water of hydration**. The hydrated salts are hydrated ionic compounds, that is molecules of water are associated with the ions of the compound. The beautiful green nickel(II) salt , for example has a formula of $NiCl_2.6\ H_2O$. The "period" between $NiCl_2$ and $6\ H_2O$ indicates that six moles of water are associated with every mole of $NiCl_2$, so the molar mass of the compound is 237.7 g/mol. Frequently, hydrated salts are obtained as crystalline salts upon evaporation of water from aqueous solutions of their salts. Common examples include $FeCl_3.6\ H_2O$ [Iron(III)chloride hexahydrate], and $CuSO_4.5\ H_2O$ [copper(II)sulfate pentahydrate]. The $FeCl_3.6\ H_2O$ consists of $Fe(H_2O)_6{}^{3+}$ and Cl^{-1} ions; the $CuSO_4.5\ H_2O$ consists of $Cu(H_2O)_4{}^{2+}$ and $SO_4(H_2O)^{2-}$ ions. Water molecules can also occur in positions in the crystal lattice that are not specifically associated with either a cation or an anion. $BaCl_2.2\ H_2O$ is an example. Most hydrates have definite compositions, containing definite number of moles of water for each mole of the anhydrous salt (hydrate salt without the attached water molecules) . Hydrates are named by naming the compound, and then naming the number of attached water molecules.The number of water molecules is indicated by using the prefixes, mono(one), di(two), tri (three) , tetra (four), penta (five), etc. e.g. $CuSO_4.5H_2O$ is called Copper (II) sulfate pentahydrate. A typical equation for the heating of a hydrate is shown below:

Hydrate -----------> Anhydrous Salt + Water

$CuSO_4.5H_2O \xrightarrow{\Delta} CuSO_4 + 5H_2O$

When $NiSO_4.6H_2O$ is heated, it loses the water of hydration, however this is also an example of a reaction that is reversible, which means that if water is added to the anhydrous compound $NiSO_4$ it will reform the hydrate compound $NiSO_4.6H_2O$. The equation below is indicative of this kind of salt.

$NiSO_4.6H_2O \overset{\Delta}{\Longleftrightarrow} NiSO_4 + 6H_2O$ The double arrow indicates reversibility.

Another example:

$CoCl_2.6H_2O \overset{\Delta}{\Longleftrightarrow} CoCl_2 + 6H_2O$ The double arrow indicates reversibility.

red — sky blue

Substances that absorb water readily are often used as dessicants which means that they attract the water molecule to them and therefore can remove water from substances. If the amount of water that is taken up is large, the substance may dissolve in the water. Substances which absorb water from the atmosphere are said to be **hygroscopic (deliquescent)** . They deliquesce usually in atmosphere of high humidity. Some other substances seem to lose large amounts of water molecules spontaneously when placed in an environment of low relative humidity (water content). These substances are said to effloresce and therefore undergo **efflorescence.** The hydrate appears powdery if squeezed.

Hundreds of hydrated salts are known. Since there is no simple way to predict how much water will be involved, it must be determined experimentally. Such an experiment usually involves heating the hydrate so that all the water is released from the crystal and driven away. Only the anhydrous salt, a material :without water," is left. The difference in the mass of the hydrated and the anhydrous material is the mass of water lost by the hydrated compound. As outlined in the following example, you can then calculate the ratio of moles of water to moles of anhydrous salt.

$BaCl_2.XH_2O$ -- Δ ---> $BaCl_2$ + XH_2O , $BaCl_2$ has a formula weight of 208.2g
Mass of hydrate = 2.00g
Mass of Anhydrous Compound was found after heating = 1.71g
Mass of water = 2.00g- 1.71g = 0.29g

Moles of $BaCl_2$ = 1.71g x (1mole/ 208.2g)= 0.0082 moles
Moles of water H_2O= 0.29g x (1mole/ 18g) = 0.0161 moles

Mole Ratio: Anhydrous $BaCl_2$: $BaCl_2$
Divide both by the smaller value

For $BaCl_2$ 0.0082 / 0.0082 = 1 For H_2O 0.0161 / 0.0082 = 1.96

Therefore the whole number ratio is : $BaCl_2 . 2H_2O$, 1:2

EXPERIMENTAL PROCEDURE

1. Obtain a sample of the hydrate and the name of its anhydrous salt from your instructor.

2. Heat a clean crucible and its cover for several minutes (10 minutes).

3. Let it cool to room temperature (Caution 1).

4. Weigh the crucible and its cover to the nearest 0.01 g.

5. Add 2.0 to 3.0 g of your sample to the crucible , reweigh and record the mass.

6. Place the covered crucible on a wire triangle and heat. First the heating should be with a low flame, and then gradually raise the temperature of the flame (Tip 1).

7. Heat for about 15 minutes after the bottom of the crucible has turned red. With the cover on it let the crucible cool down to the room temperature.

8. Weigh and record the mass.

9. Repeat the above heating, cooling, and weighing procedure until you obtain two masses within 0.05 g of each other. Your final mass must be recorded and initialed by your instructor while the crucible is on the balance.

From the mass of the original hydrate and the mass of the anhydrous salt, determine the amount of water present in the hydrate. Then determine the number of moles of anhydrous salt and the number of moles of water formed, and give the formula of the hydrate.

Tip 1: There are several ways you can start the heating procedure. You can gently move the flame under the crucible and then heat under a steady flame. You could lower the flame height and then raise it gradually. Check with your instructor before you proceed. Heating too strongly in the beginning will ruin the experiment. Why?

Safety Precaution: Exercise care when handling the hot crucible.
Handle using tongs only. Do not touch

EXPERIMENT 4 FORMULA OF A HYDRATE

DATA AND REPORT SHEET

NAME______________________________ ASSIGNED #__________ DATE ________

Name of the Anhydrous Salt:______________________Formula of anhydrous salt_______
Gram formula weight of anhydrous salt_________________

Mass of the crucible,cover and hydrate:	________	________
Mass of the crucible and cover	________	________
Mass of the Hydrate	________	________
Mass of the crucible,cover and anhydrous compound after		
1st heating	________	________
2nd heating	________	________
3rd heating	________	________
Mass of the anhydrous salt	________	________
Mass of the water present in the hydrate:	________	________

1. Number of moles of anhydrous salt in the sample:

2. Number of moles of water in the sample:

3. Number of moles of water that would combine with one mole of the anhydrous salt.

4. Formula of the hydrate:

Pre-Laboratory Assignment Experiment 4
Formula of a Hydrate
Name_________________________________ #_________ Score____

1. Compounds containing ________________ are called hydrates.
   a. hydrogen atoms b. chemically combined water c. physically bound hydrogen molecules.

2. The process when a hydrate decomposes in air to give anhydrous salt is called:
   a. efflorescence b. deliquescence c. hydration

3. In an experiment the weight of the crucible, cover and the sample was 100.45 grams. The weight of the crucible, cover and the anhydrous sample was 99.42 grams. Calculate the moles of water in the sample.

4. In the procedure section you will read "Repeat the heating, cooling, and weighing procedure until you obtain two masses within 0.05 g of each other.". What is the significance of this sentence? Choose the best answer.
   a. Improper heating and cooling will decompose the compound.
   b. All the water must be evaporated.
   c. Sample weight should not change too much in the experiment.

5. Formula of a hydrate is $A.xH_2O$. You have to find out the value of x. In an experiment you weighed out 1.023 g of $A.xH_2O$. After you follow the procedure given above, the mass of A (anhydrous salt), was found to be 0.654 g. What is the value of x? Given the molecular weight of A is 159.6 g.
   a. Mass of water released ?

   b. Moles of water driven off?

   c. Moles of A?

   d. Mole ratio of A to H_2O?

   e. value of x?

EXPERIMENT 5

IONS IN SOLUTION:TUTORIAL AND EXPERIMENT

OBJECTIVES

a) To examine the processes leading to ionic particles in an aqueous solution
b) To write formulas of the ions in aqueous solutions from the formulas of various solutes, and general solubility rules
c) To use simple chemical tests, acid/base indicators and conductivity measurements to verify the presence of certain ions in solution
d) To write net ionic equations for reactions occurring in aqueous solution and to identify spectator ions

SUPPLIES AND EQUIPMENT

Conductivity devices (one per 4 students), Spectronic 20 spectrophotometer, red and blue litmus paper, The following solids [sugar, soluble starch, oxalic acid, tartaric acid, biphenyl, dichlorobenzene, potassium hydrogen phthalate, sodium chloride, potassium bromide, calcium acetate, magnesium oxalate, calcium carbonate, lead chloride, potassium carbonate, ammonium acetate, mercurous chloride, magnesium hydroxide, calcium hydroxide, ferric sulfate, cobalt chloride, and silver chromate]
The following liquids [absolute ethanol, glycerol, glacial acetic acid, triethylamine],
The following solutions [1M HCl, 1M NaOH, 1M NH_3, 1 M HOAc, 0.1 M $CuSO_4$, 0.1 M KOH, 0.1 M $Ba(NO_3)_2$].

WHAT YOU SHOULD KNOW BEFORE YOU BEGIN

We are all very interested in chemical reactions occurring in water. Preparing a meal, washing your clothes or taking a pill to control diabetes or high blood pressure.... all these events involve chemical change in an aqueous solution. Let's quickly review some terms you probably already know. A **solution** is a homogeneous mixture (a combination of two or more substances) which can have a variable composition. The major component of the mixture is called the **solvent** and the minor component is the **solute**. Of course in most aqueous solutions the solvent is water. Much of the rest of our discussion will concentrate on the solute. We can divide solutes into two broad classes, those which cause the water to conduct an electric current, **electrolytes,** and those which do not, **nonelectrolytes**. We have always found that electrolyte solutions contain dissolved ions. Experimentally there are many ways to detect ions in solution but measuring the electrical conductivity of a solution is one of the best. What kind of solutes produce ions in solution? We'll examine three general classes of compounds, acids , bases and salts.

Acids

Unlike salts most acids are not solids, but instead are gases and liquids yet when they are dissolved in water they produce an electrolyte solution. One of the ions always present in significant amounts in an acid solution is the hydronium ion, H_3O^+. This ion is responsible for all the solution properties associated with acids, turning blue litmus paper red, tasting sharp or tart, and neutralizing bases. How is this ion formed in solution? It forms due to the chemical reaction of the solute with water. Some examples:

a) $HCl + H_2O \text{ ----> } H_3O^+ + Cl^-$

b) $H_2SO_4 + H_2O \text{ -----> } 2\ H_3O^+ + SO_4^{-2}$

In these examples the solutes donate a proton to water to form the hydronium ion. If the solute produces many hydronium ions (i.e. it is almost completely ionized) in solution it is said to be a strong acid, if it produces few ions (only partially ionized) it is a weak acid. HCl, HNO_3, H_2SO_4, $HClO_4$, HI, and HBr are common strong acids you should memorize. Most of the other acids you encounter are likely to be weak acids. Incidentally, you can often recognize acids from their formulas because they always have hydrogen written first.

Bases

When bases are dissolved in water, again the resulting solution conducts an electric current indicating the presence of ions. A special ion, the hydroxide ion, OH^-, is always present in significant amounts in the solution. This ion is formed by the chemical reaction of the base solute with the water.

c) NaOH(1) + H-O-H ----> Na^+ + H(1)-O-H + OH^-

d) NH_3 + H_2O ------> NH_4^+ + OH^-

Equation c) emphasizes the fact that the hydroxide ion in solution is not the same hydroxide ion which was a part of the solid NaOH salt, but was formed when the salt (the hydroxide ion part) accepted a proton from water. In general bases produce OH^- in water by accepting protons from water. The properties of a basic solution, turning red litmus blue, tasting bitter, feeling slippery, and neutralizing acids, are all due to the presence of the hydroxide ion in solution. You can recognize most bases because they have the OH group as part of their formula. NaOH, LiOH, KOH, $Ba(OH)_2$, $Ca(OH)_2$, are all strong bases because they produce many hydroxide ions in solution. This is partially due to the fact that they are very water soluble. $Mg(OH)_2$, for example, accepts a proton very well from water but is is not very soluble in water and hence produces only few hydroxide ions in solution. It and most other bases are weak bases.

Salts

All salts are solids which consist of closely- packed, oppositely-charged ions strongly attracted to each other by electrostatic forces. The closeness of the ions and their three dimensional arrangement in space maximizes these attractive forces. A model of a typical salt, NaCl, appears in Figure 5.1 below.

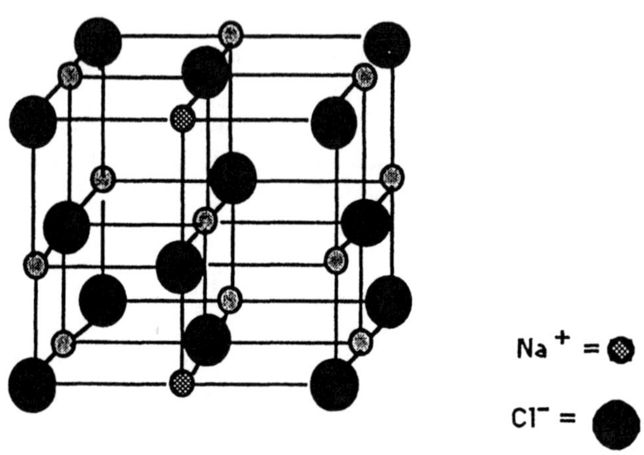

Fig. 1 Sodium Chloride Crystal Structure , an MX Type

Additionally your instructor has on display a model of a portion of solid NaCl.
NaCl has a very high melting point, 800°C, indicative of the strong forces holding the salt together. If the attractive forces are so strong in NaCl why does it dissolve so readily and easily in water because during the dissolving process (dissolution) the ions obviously must be separated from each other and uniformly spread throughout the water? The answer lies in a consideration of the energy changes that accompany dissolution of a salt. First the salt ions must be pulled apart and separated. This, of course requires energy (we say it is an endothermic process) but once the ions are separated they can form strong attractions with something else, in this case the water molecules of the solvent. The attraction of the separated ions for the water molecules is an energy liberating process (exothermic). When we add the energy changes associated with both parts of the dissolution process we often find that the net energy change is very close to zero. In other words, energetically speaking, ions are just as happy in aqueous solution as they are in the solid salt crystal. The next time you dissolve a salt or any solute in water see if you can detect a temperature change in the water. If the temperature drops, dissolution is endothermic and the energy needed for the process comes from the heat contained in the water thus cooling the water down. If the temperature rises during dissolution, the process is exothermic and heat is liberated to the water causing it to warm up.

Of course not all salts dissolve in water to the same extent. Some salts are virtually insoluble in water. We have several ways of indicating the amount of a solute which will dissolve in a given quantity of a solvent. Qualitatively, we describe solutes as soluble, partially soluble, or insoluble. So you might hear NaCl described as water soluble, but AgCl (silver chloride) described as water insoluble. If we want to be quantitative we determine the mass of the solute dissolved in a given mass of solvent and often report the results as grams of solute per 100 grams of solvent. The reasons for differing solubilities of salts involve the energy changes mentioned above but other factors play a role also. One additional factor is the favorable entropy change accompanying dissolution. As the ions are dispersed they become more disordered and this disordering process helps to lower the total energy required for dissolution. What we generally do is to develop some general solubility rules which apply to groups of salts. We group the salts on the basis of the ions that compose them.

To become familiar with the formulas and names of the common ions composing salt requires **memorization.** You should not be discouraged however because we have done a lot to systematize the memorization. Further we know that the memorization is greatly aided by the writing and naming of salt formulas... i.e., by practice. We have provided some practice in this experiment. Other practice in this important aspect of chemistry can be done on the computers in Rm 305. The attached "Table of Ions" gives the names and formulas of most of the common ions. The poem below is one attempt to aid you in your mastery of the solubility rules of the most common groups of salts.

Rules for Solubility

Potassium, sodium and ammonium salts, Whatever they may be
Can always be depended on for solubility.
When asked about the nitrates, the answers always clear,
They each and all are soluble is all we want to hear.
Most every chloride's soluble, At least we've always read,
Save silver, mercurous mercury, and (slightly) chloride of lead.
Every single sulfate is soluble 'tis said,
'Cept barium and calcium and strontium and lead.
Hydroxides of metals won't dissolve, That is all but three,
potassium, sodium and ammonium dissolve quite readily.
And then you must remember That you must not forgit
calcium and barium dissolve a little bit.
Metallic sulfides will not mix With water it is true,
But if you read the two above you'll find out which ones do.
The carbonates, insoluble, It's lucky that it's so,
Or else our marble building would melt away like snow.
(Source: Chem 13 News, February, 1989, p 10.)

Net ionic reactions

If we pour solid NaCl on top of solid $AgNO_3$ no reaction at all occurs. Yet when both these salts are dissolved in water and mixed together a reaction occurs instantaneously with the formation of a white insoluble precipitate. In solution, of course, the ions are free to move around and react with each other, but exactly which ones of the starting ions are reacting to form the white precipitate? Let's do some bookkeeping.

NaCl + $AgNO_3$ produce the following ions in water;

Na^+ , Cl^-, Ag^+, and NO_3^-

Combinations of the sodium ion and the chloride ion, and the silver ion and the nitrate ion could not produce the white precipitate because we know those combination of ions (in their salts) are soluble. That leaves two possibilities, the silver ion-chloride ion combination and the sodium ion-nitrate ion combination. Now we must use our solubility rules. All sodium salts are soluble so the precipitate could not be the sodium ion-nitrate ion combination ($NaNO_3$) . Consulting the solubility rules again indicates that the silver ion- chloride ion combination is insoluble (AgCl). That must be the white precipitate. It appears that the sodium ions and the nitrate ions do not react but are **spectator ions.** The net or overall reaction involving ions (the net ionic equation) is

$$Ag^+ + Cl^- \text{------>} AgCl .$$

Being able to write the net ionic equation is very important because it allows us to concentrate on the chemical species actually involved in the chemical change. In summary, to write the net ionic equation we had to

-write and balance the equation showing the starting salt reactants and the resulting salt (or nonsalt) products.

- use the solubility rules to determine whether a salt product is soluble or insoluble. If the salt is soluble we write the ions it produces in solution, if it is insoluble we just write the formula of the salt.

- if a reactant or product is an acid or a base we decide whether it is strong or weak. If it is strong we write the ions that it produces, if it is weak we just write the formula of the acid or base.

- We cross out or eliminate from the equation all ions which appear the same on both sides of the equation. An example appears below.

$$2\,NH_4Cl_{(aq)} + Ba(OH)_{2\,(aq)} \text{------>} BaCl_{2(aq)} + 2\,NH_{3(aq)} + 2\,H_2O$$

$$2\,NH_4^+ + 2\,Cl^- + Ba^{+2} + 2\,OH^- \text{----->} 2\,Cl^- + Ba^{+2} + 2\,NH_{3(aq)} + 2\,H_2O$$

$$2\,NH_4^+ + 2\,OH^- \text{---->} 2\,NH_{3(aq)} + 2\,H_2O$$

EXPERIMENTAL PROCEDURE

A. FORMULA WRITING AND NAMING

Complete the Table in Part A of the Report Sheet by writing the correct formula of the compound composed of the two indicated ions. When you have written the formula write a correct name for the compound

B. WATER SOLUBILITY OF SOME SUBSTANCES.

1. Your instructor has placed 20 to 25 solids (and several liquids) on the reagent table and grouped these compounds into five groups of five solids (including some liquids) each.

2. Select one group of five compounds and determine the solubility of each of the five compounds.

3. To each of five labeled test tubes containing 1 mL of deionized water, add an amount of the solid about the size of a pea(or about two drops of a liquid). Stir and observe whether the compounds dissolve. Report your results on the Report Sheet.

4. For all of the soluble compounds, write the formula of the ions in solution.

5. Your instructor has placed a table on the blackboard with two columns, soluble solids and insoluble solids. Go to the board and write the formula of all five compounds you tested in the appropriate column. Good! Now when everyone has done this examine the list together and enter the results on the Report Sheet. Do you see any patterns? If so formulate at least five solubility rules from the observations made by the entire class.

C. ELECTRICAL CONDUCTIVITY OF SOLUTES

1. Select the same five substances you used in part A.

2. For all substances which were soluble prepare a solution by mixing 1.5 g of the solid (or 10 drops of a liquid) with 30 mL of deionized water. Mix well and pour this solution into an evaporating dish.

3. Place the two wire electrodes of the conductivity device (your instructor will show you how to use it) into the solution and determine whether the LED (light emitting diode) or light bulb glows brightly, dimly, or not at all.

4. **Before discarding the solutions** test each with red and then blue litmus paper. Record your results on the Report Sheet. Be sure to rinse the electrodes with deionized water after each conductivity measurement and before switching to a new solution.

5. In addition to the five substances, measure the conductivity and acidity/basicity of 30 mL of the following solutions ; 1M HCl, 1M NaOH, 1M NH_3, 1M HOAc, 0.01 M HCl and 0.001 M HCl. It is up to you to prepare the final two HCl solutions.

6. Again your instructor has made a table on the board with the following columns, nonelectrolytes, weak acids, weak bases, strong acids, strong bases. Based on your observations place the formula of as many of your five substances as you can in the appropriate column. Copy this table into your Report Sheet.

D. NET IONIC EQUATIONS

1. The copper ion in solution is blue and absorbs light at a wavelength of 600 nanometers (nm). The concentration of copper ion in solution is directly proportional to the absorbance of a copper ion solution. Obtain 5 mL of 0.1 M $CuSO4$, add 10 mL of deionized water to it and using a cuvette measure the solution's absorbance (A) at 600 nm. Use water as a blank solution to set the Spectronic 20 to 0.000 A (Your instructor will show you how to operate the Spectronic 20).

2. Place 10 mL of 0.1M KOH in a test tube and add the 5 ml of 0.1 M copper sulfate solution to it. What is the precipitate formed?

3. Centrifuge the solution from #2, then carefully decant the aqueous layer (supernatant) away from the precipitate. Place this solution in a cuvette and determine its absorbance. Calculate the copper ion concentration of this latter solution using the equation;

$$\frac{\text{A (of 0.1M copper ion)}}{0.1\text{ M}} = \frac{\text{A (of supernatant)}}{\text{conc. of } Cu^{+2} \text{ in the supernatant}}$$

4. Other reaction: Instructor suggested.

Precaution/s The conductivity devices operate on 120 V AC. Do not touch the electrodes once the device has been plugged in.

References

Robert Eliason, Stuart Galstad, "A Simple Conductivity Device", Chem 13 News, # 213, May 1992, p.11.
Frank J. Gadek, "Easily made Electronic Device for Conductivity Measurements", J. Chem. Ed. 64, (7), 629 (1987).

EXPERIMENT 5 IONS IN SOLUTION:TUTORIAL AND EXPERIMENT

DATA AND LAB REPORT SHEET

NAME________________________________ASSIGNED #______ DATE_______SCORE______

A. FORMULA WRITING

Anions

Cations	Cl^{-1}	NO_3^{-1}	PO_4^{-3}	SO_4^{-2}	OH^{-1}	CO_3^{-2}
H^{+1}	1	2	3	4	5	6
NH_4^{+1}	7	8	9	10	11	12
Ba^{+2}	13	14	15	16	17	18
Fe^{+3}	19	20	21	22	23	24
Cu^{+2}	25	26	27	28	29	30
Hg^{+2}	31	32	33	34	35	36

Write names of all the compounds in the table above.

1______________________ 2______________________ 3______________________

4______________________ 5______________________ 6______________________

7______________________ 8______________________ 9______________________

10_____________________ 11_____________________ 12_____________________

13_____________________ 14_____________________ 15_____________________

16_____________________ 17_____________________ 18_____________________

19_____________________ 20_____________________ 21_____________________

22_____________________ 23_____________________ 24_____________________

25_____________________ 26_____________________ 27_____________________

28_____________________ 29_____________________ 30_____________________

31_____________________ 32_____________________ 33_____________________

34_____________________ 35_____________________ 36_____________________

B. SOLUBILITIES

Name	Solid	Solubility (+/-)	Ions in Solution	Ions inSolution	
	Formula		Cations	Anions	
1					
2					
3					
4					
5					

Develop five solubility rules based on the class observations.

____

____

____

____

____

C. CONDUCTIVITY TESTS

Solute	Brightness of LED	Litmus (R--> B)	Electrolyte	Ions in Solution
Formula	or Bulb (Bright ,Dim or Off)	or (B--> R)	(Strong, Weak,	
			or Non)	
1				
2				
3				
4				
5				
1M HCl				
1M NaOH				
1M NH_3				
1HOAc				
0.01M HCl				
0.001M HCl				

C. (Cont.)

Which solutes are not electrolytes?____

Examine all the HCl results. What did you observe as the concentration of HCl decreased? Explain this in terms of the number and type of ions in solution.

____

____

____

Table from Blackboard

Nonelectrolyte	Weak Acid	Weak Base	Strong acid	Strong Base

D. NET IONIC EQUATIONS

Equation for $CuSO_4$ + KOH----------->

Net ionic equation____

What is the precipitate?Name__________________________ Formula______________

Initial A__________________________ Initial $[Cu^{+2}]$____________________

Final A__________________________ Final $[Cu^{+2}]$____________________

How did the concentration of Cu^{+2} in the solution change? What caused this change?

The Spectator ions in this reaction are ________ Explain

Other Suggested Reactions:

TABLE OF IONS

CATIONS		ANIONS	
Aluminum ion -------------	Al^{+3}	acetate--------	$OAc^{-}/Ac^{-}/C_2H_3O_2^{-1}$
Ammonium ----------------	$NH4^{+1}$	bicarbonate----	HCO_3^{-1}
Antimony (III) or Antimonous	Sb^{+3}	carbonate------	CO_3^{-2}
Antimony (V) or Antimonic--	Sb^{+5}	cyanide--------	CN^{-1}
Arsenic (III) or Arsenious---	As^{+3}	cyanate-------	CNO^{-1}
Arsenic (V) or Arsenic -----	As^{+5}	acetylide-----	C_2^{-2}
Barium ion ----------------	Ba^{+2}	thiocyanate----	SCN^{-1}
Bismuth ion ---------------	Bi^{+3}	dithioorthoarsenate	$AsO_2S_2^{-3}$
Cadmium ion -------------	Cd^{+2}	oxalate--------	$C_2O_4^{-2}$
Calcium ion-----------------	Ca^{+2}	chloride-------	Cl^{-1}
Chromium (II)or Chromous ion	Cr^{+2}	hypochlorite----	ClO^{-1}
Chromium (III) or Chromic ion	Cr^{+3}	chlorite--------	ClO_2^{-1}
Copper(I) or Cuprous ion ----	Cu^{+1}	chlorate--------	ClO_3^{-1}
Copper(II) or Cupric ion -----	Cu^{+2}	perchlorate-----	ClO_4^{-1}
Hydrogen ion --------------	H^{+1}	(ortho)arsenate-	AsO_4^{-3}
Hydronium ion -------------	H_3O^{+1}	(meta)arsenate--	AsO_3^{-1}
Iron (II) or Ferrous ion ------	Fe^{+2}	(ortho)arsenite--	AsO_3^{-3}
Iron (III) or Ferric ion ------	Fe^{+3}	(meta)arsenite--	AsO_2^{-1}
Lead(II) or Plumbous ion -----	Pb^{+2}	chromate-------	CrO_4^{-2}
Lead (IV) or Plumbic ion -----	Pb^{+4}	dichromate-----	$Cr_2O_7^{-2}$
Lithium ion ----------------	Li^{+1}	(ortho)chromite-	CrO_{3-}^{3}
Magnesium ion -------------	Mg^{+2}	(meta) chromite-	CrO_2^{-1}
Manganese(II),Manganous ion	Mn^{+2}	(ortho)borate----	BO_3^{-3}
Mercury (I) or Mercurous ion	Hg_2^{+2}	(meta)borate-----	BO_2^{-1}
Mercury (II) or Mercuric ion-	Hg^{+2}	nitride-----------	N^{-3}
Nickel ion -----------------	Ni^{+2}	amide------------	NH_2^{-1}
Potassium ion -------------	K^{+1}	imide------------	NH^{-2}
Silver ion -----------------	Ag^{+1}	azide------------	N_3^{-1}
Sodium ion ----------------	Na^{+1}	peroxide---------	O_2^{-2}
Strontium ion -------------	Sr^{+2}	superoxide-------	O_2^{-1}
Thallium(I) or Thallous ion---	Tl^{+1}	oxide------------	O^{-2}
Tin(II) or Stannous ion ------	Sn^{+2}	hydroxide--------	OH^{-1}
Tin (IV) or Stannic ion ------	Sn^{+4}	hydride----------	H^{-1}
Zinc ion ------------------	Zn^{+2}	orthophosphate---	PO_4^{-3}
Cobalt (II) or Cobaltous ion --	Co^{+2}	monothiophosphate-	PSO_3^{-3}
Cobalt(III) or Cobaltic ion ---	Co^{+3}	metaphosphate----	PO_3^{-1}
Uranyl ion ----------------	UO_2^{+2}	pyrophosphate----	$P_2O_7^{-4}$
Vanadyl ion --------------	VO^{+2}	orthosilicate------	SiO_4^{-4}
Vanadium ion -------------	V^{+5}	metasilicate------	SiO_3^{-2}

ANIONS CONTINUED

sulfide ion--------	S^{-2}
bisulfide----------	HS^{-1}
sulfate-----------	SO_4^{-2}
bisulfate---------	HSO_4^{-1}
bisulfite---------	HSO_3^{-1}
sulfite ----------	SO_3^{-2}
thiosulfate------	$S_2O_3^{-2}$
pyrosulfate------	$S_2O_7^{-2}$
manganate-------	MnO_4^{-2}
permanganate----	MnO_4^{-1}
nitrate----------	NO_3^{-1}
nitrite----------	NO_2^{-1}
ferricyanide or hexacyanoferrate(III)	$Fe(CN)_6^{-3}$
ferrocyanide or hexacyanoferrate(II)	$Fe(CN)_6^{-4}$

Pre-Laboratory Assignment Experiment 5
Ions in Solution: Tutorial and Experiment Name______________________#______Score_____

1. Using the 100 and 10 mL graduated cylinders in your drawer, give detailed instructions for the preparation of 30 mL of 0.01M and 0.001 M HCl solutions from 0.1M HCl.

2. Write a short poem (or a rap) describing some of the important properties of acids and bases.

3. A 0.12 M solution of copper ion in water has an absorbance of 0.568 at 600 nm. After reaction of the solution with sulfide ion the absorbance of the solution is 0.102. Calculate the copper ion concentration remaining in the solution after precipitation of copper sulfide.

4. Write the formulas of all the salts which could possibly be formed from various combinations of the following ions;
   Na^{+} , Cl^{-}, Ag^{+} , SO_4^{-2} and NO_3^{-}

   Name each of these compounds.

EXPERIMENT 6

CHEMICAL REACTIONS

OBJECTIVES

a) To become familiar with the types of chemical reactions
b) To study five kinds of chemical reactions
c) To predict reaction products
d) To complete and balance molecular, ionic and net ionic equations
e) To become familiar with solubility rules of substances

SUPPLIES AND EQUIPMENT

15 Test tubes, Test tube holder, Bunsen Burner, Lighter, Crucible, Solids (Cu strips 1/4 inch, $CuSO_4.5H_2O$, $KClO_3$, Mg Ribbon-1/2 strips), 1M aqueous solutions of ($NiCl_2$, $BaCl_2$, Na_3PO_4, NaOH , HNO_3, H_2SO_4, HCl, NH_4OH), O.1M $AgNO_3$, 6M HCl, Conc. NH_4OH,(15M), Conc. HCl (12M)

WHAT YOU SHOULD KNOW BEFORE YOU BEGIN

Chemical reactions (ordinary) can be classified into five kinds. Chemical equations are the chemist's shorthand notations for reactions. One kind of reaction is that of the **single displacement**
(single replacement) . In this kind of reaction, one element displaces another from a compound. The displacement occurs due to the greater activity of the element as compared to the element in the compound.

$$Zn + CuCl_2 \text{--->} ZnCl_2 + Cu$$

Therefore Zinc, (Zn) is more reactive than Copper, (Cu).
Another kind is that of **combination.** In this kind of reaction, two or more reactants are combined together to form one compound.

$$H_2 + Br_2 \text{--->} 2HBr$$

A third kind is that of **decomposition.** In this kind of reaction, one compound is broken down into simpler substances. Heat is usually associated with these kinds of reactions. The heat serves to cause the decomposition.

$$2KClO_3 \xrightarrow{\Delta} 2KCl + 3O_2$$

A fourth kind of reaction is that of **Metathetical (Double Displacement)** . In this kind of reaction, there is an exchange of two elements . There is a double switching of reactant partners. Included in this kind is the reaction of salts with acids, salts with salts, salts with bases, and acids with bases (called neutralization).

$$\underset{\text{salt}}{AgNO_3} + \underset{\text{acid}}{HCl} \text{----------->} \underset{\text{salt}}{AgCl} + \underset{\text{acid}}{HNO_3}$$

$$\underset{\text{salt}}{Pb(NO_3)_2} + \underset{\text{salt}}{K_2CrO_4} \text{--->} \underset{\text{salt}}{PbCrO_4} + \underset{\text{salt}}{2KNO_3}$$

$$\underset{\text{salt}}{2KCl} + \underset{\text{base}}{Mg(OH)_2} \text{--->} \underset{\text{base}}{2KOH} + \underset{\text{salt}}{MgCl_2}$$

The neutralization involves H^{+1} from acids, and OH^{-1} from bases. When classical acids react with classical bases, salts and water are formed.

HCl	+ NaOH------>	NaCl	+	HOH
acid	base	salt		water

The cations in each of the compounds switched partners with the anions of the other compound.

A fifth kind of reaction is that of **Redox** . This kind of reaction involves a change in oxidation number of some elements. Single displacement, decomposition, and combination reactions also can be characterized as redox reactions.

Single displacement : $Zn + CuCl_2 \text{ ---> } ZnCl_2 + Cu$

The oxidation state of Zn is 0 as an element , and +2 in $ZnCl_2$, The oxidation state of Cu is +2 in $CuCl_2$, and 0 as an element, Cu.

Combination : $H_2 + Br_2 \text{ ---> } 2HBr$

The oxidation state of H in H_2 is 0, and (+1 in HBr), The oxidation state of Br in Br_2 is 0, and -1 in HBr.

Decomposition : $2HgO(s) \xrightarrow{\Delta} 2Hg(l) + O_2$

The oxidation state of Hg in HgO is +2 , and 0 in Hg(l).

Another kind of reaction is that of **combustion**. These reactions are also examples of **Redox** reactions. Most of the combustion reactions are reactions that occur when organic hydrocarbons are ignited in air (excess oxygen) with the formation of water and carbon dioxide in complete combustion or of water and carbon monoxide being formed in incomplete combustion (oxygen limited). Examples of each are :

$C_2H_4(g) + 3O_2(g) \text{ ----> } 2CO_2(g) + 2H_2O(l)$ complete combustion

$2CH_4(g) + 3O_2(g) \text{ ----> } 2CO(g) + 4H_2O(l)$ incomplete combustion

Another type combustion is : $2Mg + O_2 \text{--> } 2MgO$.

In this experiment, the various kinds of reactions will be investigated. In observing chemical reactions, the following should be noted: heat transfer (temperature changes), solid formation (precipitation), gas evolution , and color changes. The designations of the aforementioned chemical behavior is summarized below:

solid--------------->(s),	___(underline),	(down arrow, ↓) (evidenced by cloudiness)
gas----------------->	(g),	up arrow, ↑ (evidenced by bubbling in solution)
liquid-------------->	(l)	
heat --------------->	Δ	
water (aqueous)->	(aq)	indicates a substance is soluble in water.

Also, Metals can be indicated with valences, e.g. Cu^o, Fe^o, Zn^o. This is useful in balancing ionic equations. $2Ag^{+1} + Cu^o \text{ ---------> } 2Ag^o + Cu^{+2}$

To facilitate the writing of equations, the names of reactants and products are given using the above:

$Zn(s)$ +	$CuCl_{2(aq)}$ ---------------->	$ZnCl_{2(aq)}$	+ $Cu(s)$	
Zinc (s) +	Copper (II) chloride	Zinc chloride (aq)	+ Copper(s)	
$H_{2(g)}$ +	Br_2 (g) --------->	2HBr(g)		
Hydrogen	Bromine	Hydrogen bromide		
2HgO(s) $\xrightarrow{\Delta}$	2Hg(l) +	O_2		
Mercuric oxide	Mercury metal	Oxygen gas		
$AgNO_{3(aq)}$ +	HCl (aq) ------>	AgCl(s) +	HNO_3(aq)	
Silver nitrate	Hydrochloric acid	Silver chloride	Hydrogen nitrate	
$Pb(NO_3)_2$(aq) +	K_2CrO_4 (aq) ---->	$PbCrO_4$(aq) +	$2KNO_3$(aq)	
Lead (II) nitrate	Potassium chromate	Lead (II) chromate(s)	Potassium nitrate (aq)	
HCl(aq) +	NaOH(aq) ------>	NaCl (aq) +	HOH(l)	
Hydrochloric acid	Sodium hydroxide	Sodium chloride	Water	

The chemical equations written with reactants and products as compounds, molecules , or elements are called **molecular** equations. When the substances are written as they exist in the various states the equations become **ionic** equations. In general all substances are written as ions in aqueous medium with the exception of insoluble solids , liquids and gases. The exception to the general rule occurs in weak electrolyte reactions. When the equations include only those substances which have undergone changes , they are called **net ionic** equations.

Examples of which are shown below:

Zn (s) + $CuCl_2$ (aq)--------> $ZnCl_{2(aq)}$ + Cu (s) **Molecular**

Zn(s) + Cu^{+2} (aq) + 2 Cl^{-1}(aq)-----> Zn^{+2}(aq)+ 2Cl^{-1}(aq) + Cu(s) **Ionic**

Zn(s) + Cu^{+2}(aq) ----> Zn^{+2}(aq) + Cu(s) **Net Ionic**

The 2 Cl^{-1}(aq) are called "spectator ions" and are not of real concern in the reaction.

$AgNO_{3(aq)}$ + HCl (aq) ----------> AgCl(s) + HNO_3(aq) **Molecular**

Ag^{+1} (aq)+ $NO_3^{-1}{}_{(aq)}$ + H^{+1}(aq) + Cl^{-1}(aq)-->AgCl(s) + H^{+1}(aq) + NO_3^{-1}(aq) **Ionic**

Ag^{+1} (aq)+ Cl^{-1}(aq)------------> AgCl(s) **Net ionic**

The $NO_3^{-1}{}_{(aq)}$ + H^{+1}(aq) ions are the "spectator ions" and are not of real concern in the reaction.

HCl(aq) + NaOH(aq) ---> NaCl(aq) + HOH(l) **Molecular**

H^{+1} (aq)+ $Cl^{-1}{}_{(aq)}$ + Na^{+1}(aq) + OH^{-1}(aq)--->Na^{+1}(aq) + Cl^{-1}(aq) + HOH(l) **Ionic**

H^{+1}(aq) + OH^{-1}(aq)-------------> HOH(l) **Net ionic**

Net ionic = The essential reaction of an acid with a base in this Neutralization Reaction.

$FeCO_{3(s)}$ + 2HCl (aq)--->Fe^{+2} (aq)+ 2$Cl^{-1}{}_{(aq)}$+H_2CO_3(aq) =>HOH(l) +CO_2(g)**Molecular**

$FeCO_{3(s)}$ + 2H^{+1} (aq) +2$Cl^{-1}{}_{(aq)}$ ---->Fe^{+2} (aq)+ 2$Cl^{-1}{}_{(aq)}$+HOH(l) +CO_2(g)**Ionic**

$FeCO_{3(s)}$ + 2H^{+1} (aq) --->Fe^{+2} (aq)+ HOH(l) + CO_2(g) **Net ionic**

EXPERIMENTAL PROCEDURE :

Note all observations on the report sheet. Carefully observe all reactions. Always note the original state before reaction and that after reaction. Note the actual color changes, evolution of gases (bubbling), formation of solids (cloudiness, precipitates), moisture accumulated (wet), heat (did the test tube warm up or cool down by feeling or by use of a thermometer) etc.

1. Ignite the bunsen flame and note all observations.

2. A. Place 0.1 g of $CuSO_4.5H_2O$ into a clean dry 200 mm Pyrex test tube. Place the test tube in a clamp set up for heating. Heat the test tube for 5 minutes and note all changes. Let the test tube cool to room temperature, add water and note what happens.

3. A. Place a 1/2 inch piece of magnesium ribbon into a test tube containing 30 drops of 6M HCl.

   B. Place a 1/4 inch square piece of polished copper metal into 5ml of 1.0M $AgNO_3$ solution. Observe for at least five minutes.

4. A. Pour 20 drops of 1M $BaCl_2$ into one test tube, and add 2ml of 1M H_2SO_4 and note your observations.

   B. Pour 20 drops of 1M $NiCl_2$, into a second test tube, and add 20drops of 1M Na_3PO_4 and note your observations.

   C. Pour 20 drops of 1M HNO_3 into a clean test tube, insert a thermometer in the test tube, and add 20 drops of 1M NaOH. Read the thermometer when the reagents are mixed.

   D. Add 10 drops of concentrated, 15M NH_4OH to a test tube, and slowly pour down the inside of the test tube 10 drops of concentrated, 12M HCl. Note all observations.

   E. Pour 5 drops of 1M K_2CrO_4, into a second test tube, and add 5 drops of 0.1M $Pb(NO_3)_2$ and note your observations.

5. A. Insert a piece of magnesium ribbon into a bunsen flame, and observe what occurs. **Do not look directly into the flame while heating.** Note and record all observations. Drop the residue into an evaporating dish containing 1mL of distilled water. Using Technique 17T on page 15, test for acidity .

   B. Place a 0.5g sample of $KClO_3$ into a hard glass test tube. Add a pinch of MnO_2, and assemble apparatus in Technique 1T on Page 7. Mount test tube onto test tube rack. Heat gently with a low flame displacing the water in the glass bottle, and carefully observe the appearance of the salt. Slide a glass plate over the mouth of the bottle while under the water. Light a splint, remove cover from bottle and **immediately** insert it at the mouth of the bottle, and note what occurs. This should be demonstrated by the lab instructor. It is considered dangerous. **INSTRUCTOR DEMONSTRATION**

   Safety Precautions:

   1. When heating a test tube containing liquids or solids, never point the test tube at anyone.
   2. Review Techniques Sections on Lighting a Burner, and Heating a test tube.

EXPERIMENT 6 CHEMICAL REACTIONS

DATA AND LAB REPORT SHEET

NAME_______________________________ASSIGNED #______ DATE____________

1. $CH_4 + O_2$

Reaction Evidence:____

Balanced Molecular Equation:

____

2A. $CuSO_4 \cdot 5H_2O + \Delta$

Reaction Evidence:____

Balanced Equation:

____

3A. Mg + HCl

Reaction Evidence:____

Balanced Equation:

____

3B. Cu + $AgNO_3$

Reaction Evidence:____

Balanced Equation:

____

4A. $BaCl_2 + H_2SO_4$

Reaction Evidence:____

Balanced Molecular Equation:

____

Ionic equation | Net ionic equation

____

4B. $NiCl_2 + Na_3PO_4$

Reaction Evidence:____

Balanced Molecular Equation:

____

Ionic equation | Net ionic equation

____

4C. HNO_3 + NaOH

Reaction Evidence:____

Balanced Molecular Equation:

____

Ionic equation

Net ionic equation(essential acid-base)

4D. NH_4OH + HCl

Reaction Evidence:____

____

Balanced Molecular Equation:

____

Ionic equation

Net ionic equation

____

4E. K_2CrO_4 + $Pb(NO_3)_2$

Reaction Evidence:____

Balanced Molecular Equation:

____

Ionic equation

Net ionic equation

____

5A. Mg + O_2

Reaction Evidence:____

Balanced Equation:

____

5B. $KClO_3$ + Δ DEMO

Reaction Evidence:____

Balanced Equation:

____

Classify the above reactions as to type or kind of reaction. If any reaction can be classified in more than one category, indicate all categories that apply to each.

1__________ 2A__________3A__________4A__________4B__________ 4C__________

5A__________ 5B__________

Metathetical-M, Single Replacement-S, Redox-R, Combination-C, Decomposition-D, Acid-Base-AB, Combustion-CN

Indicate the water solubility of the following substances: insoluble (i) soluble (s) and their colors:

$Ni_3(PO_4)_2$__________ color__________ $BaSO_4$__________ color__________

AgCl__________ color__________ KCl__________ color__________

$PbCrO_4$__________ color__________

Pre-Laboratory Assignment Experiment 6
Chemical Reactions Name____#______Score _____

1. Define the following terms:
   Metathetical Reaction

   Redox Reaction

   Precipitate

   Acid-Base Neutralization

2. Using the solubility tables in the appendix, what is the qualitative water solubility of the following substances? (e.g. soluble, insoluble, partially soluble)

   $PbCl_2$__________ $Cu(NO_3)_2$____________ AgCl_________

3. What is a chemical equation?

4. If bubbling occurs in a test tube when two substance are mixed, what is this evidence of?

5. If in a test tube when two substance are mixed, the test tube warms up, what is this evidence of?

6. Using alphabet notations, write expressions for the following reaction kinds:
   single displacement combination decomposition

7. Using water, how could distinguish between $BaSO_4$ and Na_2SO_4?

EXPERIMENT 7

ACTIVITY OF METALS

OBJECTIVES

a) To experimentally verify the presence of metal ions in solution after the redox reactions of several metals
b) To determine the relative reactivity of some common metals with hydrogen ion and other metal ions
c) To determine the relativity stability of some metal oxides
d) To attempt a correlation between a metal's reactivity and its position in the Periodic Table

SUPPLIES AND EQUIPMENT

Bunsen Burner, Lighter, Pyrex Test Tubes, Calcium Curled strips, Phenolphthalein Indicator, Wooden Splints, Steel Wool, Microtrays, Ring Stand0.1M $Pb(NO_3)_2$, 0.1M $Cu(NO_3)_2$, 0.1M $Zn(NO_3)_2$, 0.1M $FeSO_4$ (Freshly prepared), 0.1M $Mg(NO_3)_2$, Metallic Solids-1/4 inch strips or pieces (Cu wire, Mg Ribbon, Fe Filings, Pb, Mossy Zn), Solid Oxides (CaO, BaO, PbO_2)

WHAT YOU SHOULD KNOW BEFORE YOU BEGIN

Most of you will either give or receive some gold or silver items (wedding ring, class ring, etc.,) in the next few years, but why gold and silver and not zinc or tin or iron? Of course you already know the reason, among the metals gold, silver and platinum are relatively unreactive. They don't easily corrode (rust) or decay, yet metals like iron and zinc decay or corrode readily. Today's experiment aims to measure the ease with which a metal "corrodes", to list the metals in order of ease of corrosion, and to see these corrosion reactions as belonging to a single larger class of chemical reactions.
The word "reactivity" or in this experiment activity, refers to how **fast** a substance (a metal) reacts with some other reagent and we can measure this by determining how much time it takes for a fixed mass of a given metal to be used up. A second way to determine reactivity is to allow the same amount of two different metals to react with an excess of some other reagent and note which of the two metals is consumed first. For example if we drop sodium metal and iron metal into water we can see the sodium metal dissolving and when it's all gone the iron metal remains. We say that, in this instance, sodium is more reactive that iron. Let's look more closely at the reaction of sodium metal and water.

As you saw when your instructor demonstrated this reaction, the sodium metal dissolved, a flammable gas was formed, and the resulting solution turned red litmus paper to blue. We can reliably interpret these observations to show that hydrogen gas and hydroxide ions are products of this reaction. To balance the reaction (among other things, to make sure that the total charge of reactants equals the total charge of the products) requires that sodium **ion** (Na^+) is also a product of this reaction even though we have no easy experimental method to show its presence.

$$2Na + 2H_2O \text{ ----> } 2Na^+_{(aq)} + 2OH^-_{(aq)} + H_{2(g)}$$

sodium water sodium ion hydroxide ion hydrogen gas

Often when most metals dissolve or react they are converted into the metal ion. This process is called **oxidation** and is accompanied by the **loss of electrons** from the metal. These electrons have to go somewhere and they always end up being accepted by some other atom. The **gain of electrons** by an atom is called **reduction**. In the above example the electrons are gained by the hydrogen atoms of the water molecule.

$2Na_{(s)} \text{ ----> } 2Na^+_{(aq)} + 2\,e^-$ (oxidation) $2H_2O + 2e^- \text{ ----> } 2OH^-_{(aq)} + H_{2(g)}$ (reduction)

The overall reaction is called an **oxidation-reduction or redox reaction**. The corrosion and decay of metals and the burning of any substance are also examples of redox reactions.

To identify a redox reaction it helps to know what the products of the reaction are. We can determine the products with a little detective work based on the interpretation of observations.
Here are some observations and the way chemists generally interpret them.

Observation	Interpretation
1. Bubbles are formed in a solution	A gaseous substance is formed
2. The gas formed is lighter than air and burns explosively when a burning wooden splint is inserted	The gas is hydrogen, H_2
3. The gas formed causes a glowing wooden splint to burst into full flame	The gas is oxygen, O_2
4. The solution conducts an electric current	Ions are present in the solution
5. The solution turns blue litmus to red	Hydrogen ion, H^+, is in the solutio, acidic
6. The solution turns red litmus to blue or phenolphthalein turns pink in the solution	Hydroxide ion, OH^-, is in the solution, basic
7. The solution turns blue after reaction	Copper ion, Cu^{+2}, is in the solution
8. A silver-colored liquid metal is formed	The liquid metal is mercury, Hg

In the case of the reaction of other metals it is sometimes possible to easily show that the metal ion is formed and is present in the solution in which the reaction took place. Copper ion in solution is a deep blue color. What chemical species do you think is formed when a copper-containing penny is dissolved in concentrated nitric acid and the solution turns blue?

There is yet a third way to measure the relative reactivity of atoms involved in a redox reaction. An element which is more reactive is able to take the place of a less reactive element that is in solution or combined with a third atom. In the reaction above sodium atoms joined the solution and took the place of hydrogen atoms. These hydrogen atoms came out of solution as hydrogen gas. Thus sodium atoms are more reactive than hydrogen atoms. In general if the reaction below occurs

$$X_{(s)} + YZ_{(aq)} \text{ ----> } XZ_{(aq)} + Y_{\text{solid or gas}}$$

we can conclude that X is more reactive than Y. Obviously the reverse reaction will not occur.

$$Y_{(s)} + XZ_{(aq)} \text{ --X--> } YZ_{(aq)} + X_{\text{solid or gas}}$$

Many of the elements react readily with oxygen, O_2, to form binary compounds in which the metal is oxidized, that is all except mercury, silver, platinum ,palladium and gold. For these "noble" metals the oxides are prepared in a series of reactions. These elements have so little desire to be oxidized that , upon heating, their metal oxides (HgO, Ag_2O, PtO, PdO, Au_2O_3) readily release oxygen gas reforming the free metal.

$$2HgO_{(s)} \text{ ----heat---> } 2Hg_{(l)} + O_{2(g)}$$

Is it possible to relate the reactivity of a metal to its location in the Periodic Table? Yes. Since many metal reactions are oxidations (electrons are lost) those metals which lose electrons readily (low ionization energies) should be very reactive. So, look for the most reactive metals at the bottom left-hand corner of the Periodic Table.

STUFF YOU CAN USE

When you buy gold items the purity of the gold may be expressed in **karats**

% Gold	100	91.7	75	58.5	41.6
Mark on Jewelry	24k	22k	18k	14k	10k

If you don't have much money you can buy solid gold substitutes;

gold filled- a layer of at least 22 k gold permanently bonded by heat and pressure to the surfaces of a supporting metal, then rolled or drawn to a prescribed thickness. The karat gold must be 1/20 by weight of the total metal content.
heavy gold electroplate- a film of fine gold deposited on other metals electrically at a thickness of at least 100 millionths of an inch.
gold electroplate- the film must be at least 7 millionths of an inch thick.
["Chem 13 News", November 1980, p15.]

What color is copper metal? Wrong!! Copper is yellow, like gold. The red-brown color is due to a coating of red Cu_2O and black CuO on the surface of the copper. [J. Chem. Ed., May 1989,p400].

EXPERIMENTAL PROCEDURE

A. CALCIUM METAL WITH WATER

1. Obtain a curled strip of calcium metal, a cork stopper. Mount the test tube in a three prong clamp on a ring stand.
2. Add 5-7mL of distilled water, add the Ca to the test tube(Pyrex) and immediately stopper (not to tightly) the test. Observe what occurs.
3. After a vigorous reactions occurs, light a wooden splint, quickly remove the stopper and place the lighted splint at the mouth of the test tube to test the flammability, note and record what occurs.
4. Once the reaction subsides, add three drops of phenolphthalein indicator to the solution in the test tube. Note and record what occurs. From your observations infer the identity of the products of the reaction and write a balanced equation.
   BE SURE YOU ARE WEARING SAFETY GLASSES.

B. METAL-METAL ION REACTIONS (GROUP ASSIGNED)

1. Obtain a microtray (96 wells) from the storeroom and , with a piece of tape, label the tray as indicated in Figure 8.1.
2. In all the wells of column one place 1 drop of 0.1M $Pb(NO_3)_2$.
3. In all the wells of column two place one drop of 0.1M $Cu(NO_3)_2$, etc., .
4. In each well of row A place a thin metal strip of Pb (cleaned with sand paper before use) about 3 mm wide and a cm long.
5. In each well of row B, place a thin strip of Cu, etc. Repeat for the other metal ions. Before using each metal be sure to clean its surface by rubbing it vigorously with sandpaper. Allow the metal to remain in contact with the solution for about 30 seconds then remove the metal and look for signs that the metal has reacted. Sometimes reaction is indicated by a darkening of the metal surface. Assume that in each well in which reaction occurs the metal has formed metal ions. For each reaction that occurred write a balanced equation labeling the element oxidized and the element reduced. After considering all your equations list the metals in order of increasing reactivity.

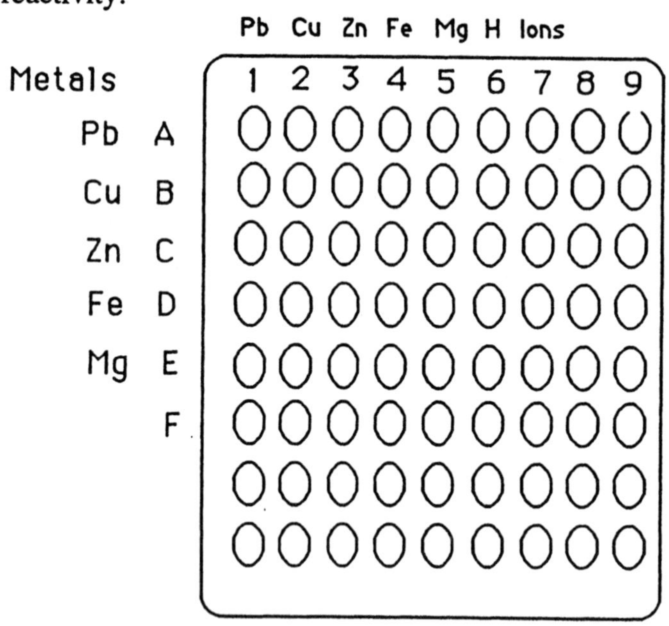

Fig. 8.1 Microtray for Reaction of Metals and Metal Ions

C. STABILITY OF METAL OXIDES

1. Obtain a pea-sized amount of the following solids, BaO (Barium Oxide) and PbO_2 (Lead (IV) oxide and CaO (calcium oxide).
2. Place each in a test tube and heat strongly with a Bunsen burner for about two minutes.
3. During this period of heating, place a glowing wood splint in the mouth of the test tube and observe. Lead compounds are hazardous solid wastes and must be disposed of properly by law. When finished with the test tubes containing the lead compound, return the test tube and contents to the lab instructor. <u>Do not toss these compounds in the waste containers or in the sink.</u> From your observations, write balanced equations describing any reactions which occurred. Was any oxide unreactive?
4. Arrange the metals, Ba, Ca and Pb in order of increasing ease of oxidation.

EXPERIMENT 7 ACTIVITY SERIES

DATA AND LAB REPORT SHEET

NAME________________________________ASSIGNED #_____ DATE______ SCORE______

A. Calcium Metal and Water

Properties of Gas Formed____

Observation of Phenolphthalein Addition____

Identity Of The Reaction Products:____

Balanced Equation ____

B. Metal-Metal Ion Reactions

	Pb^{+2}	Cu^{+2}	Zn^{+2}	Fe^{+2}	Mg^{+2}	H^{+1}
Pb						
Cu						
Zn						
Fe						
Mg						

Place an **R** in the Table for any observed reaction , and place an **NR** for no reactions.

Write ionic equations for each metal if there was a reaction:

Pb:	
Cu:	
Zn:	
Fe:	
Mg	

Order of Increasing Reactivity____

Write the net ionic equations for each metal and H^{+1} reactions:

______________________ ______________________

______________________ ______________________

____

Arrange in order of increasing reactivity. The ability to displace hydrogen ion:

____

C. Stability of Metal Oxides

Observations and Balanced Equations For Heating The Following Oxides:

Lead Oxide______________________

____

Calcium Oxide______________________

____

Barium Oxide______________________

____

Arrange the metal oxides from most stable to least. stable______________________

Pre-Laboratory Assignment Experiment 7
Activity Series Name________________________________#______Score______

1. List two American coins and the principal metal composing each coin.
   What do these metals all have in common?

2. Aluminium metal reacts with dilute hydrochloric acid according to the following equation:

$$2Al_{(s)} + 6HCl_{(aq)} \dashrightarrow 2AlCl_{3(aq)} + 3H_{2(g)}$$

a) From the equation above, what substances are dissolved in the solution before and after the reaction of aluminum with HCl?

b) Which product is a gas? Will this gas cause a glowing splint to burst into flames , react explosively with a burning wood splint or do nothing at all with a burning splint placed in the gas?

c) If one mole of $AlCl_3$ is formed, how many moles of Cl^- are present in solution?

d) Which element is oxidized? Which element is reduced?

3. Consider the following redox reactions:

$$X + Y^+_{(aq)} \dashrightarrow Y + X^+_{(aq)}$$
$$M + Y^+_{(aq)} \dashrightarrow \text{no reaction}$$
$$Z + X^+_{(aq)} \dashrightarrow Z^+_{(aq)} + X$$

Arrange the metals X, Y, M and Z in order of increasing reactivity
least reactive most reactive

-->

4. List the color of the following substances:

Cu ______________ CaO ______________ Cu^{+2} ______________

Phenolphthalein in a solution containing OH^-________

EXPERIMENT 8

SPECTROSCOPY

OBJECTIVES

a) To explain the theory of a line spectra.
b) To investigate flame emission colors for excited metal ions.
c) To use a spectroscope and determine wavelengths.
d) To investigate the hydrogen spectrum.
e) To determine a value of Planck's Constant , h.
f) To plot scientific data and determine the slope .

SUPPLIES AND EQUIPMENT

Spectroscopes, Filter paper strips, Spectrum Lamps, Spectrum Power Supply, Test tubes (Pyrex), Wire Loops, Concentrated HCl, Distilled water, Bunsen burner, Solids ($CaCl_2$, $BaCl_2$, $SrCl_2$, $CuCl_2$, LiCl, KCl, NaCl, Incandescent bulbs, Spectrophotometer, Graph Paper

WHAT YOU SHOULD KNOW BEFORE YOU BEGIN

Presently, the theory of atomic structure states that the atom can only have discrete energy values (quantized). We know that the atom has protons and neutrons in the nucleus and electrons surrounding the nucleus. The energy levels of atoms are designated by an integer, **n**, which is called the **principal quantum number.** At ambient temperature, most of the hydrogen atoms have energy that is equivalent to the **n=1 level**. The atom is said to be in the **ground state**. When an atom absorbs energy from either heat or electrical discharge, some of the atoms will absorb this energy and become **excited**. In this state, the electron will occupy an energy level where n = 2,3, etc. The atom is in an excited state, E_h. This is an Absorption of Energy process and the result yields an **Absorption Spectrum.** When the electron loses this energy and falls back to a lower level or to the ground state, E_l. This is an Emission of Energy process, and the result yields an **Emission Spectrum**. The electron emits a photon of energy, E_{photon}, when it returns to the n=1 ground state.

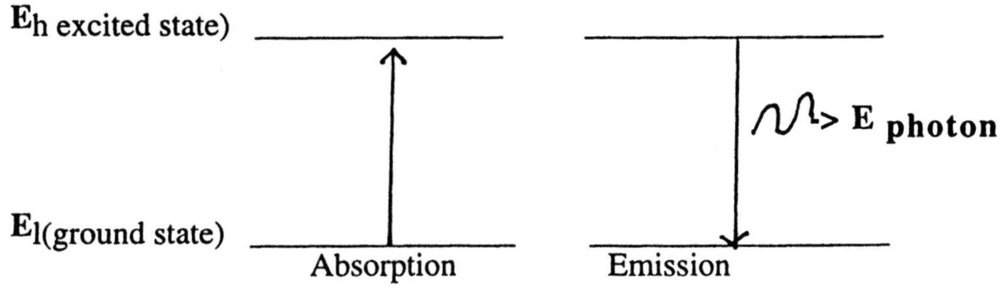

$$E_{photon} = E_h - E_l = \Delta E_{atom} = hc/\lambda = h\nu$$

When the atom loses energy, often light is given off at definite frequencies. This light is the photon which is described as a particle and a wave or neither. The proportionality constant between a photon and its energy is called Planck's Constant , h, after Max Planck. The value is 6.626E-34J.s or 6.626E-27erg.s. Therefore a hydrogen electron can lose energy by emitting a photon with an energy corresponding to the transition from an upper to a lower energy level, e.g. n =3--> n=2.

The speed of light (c) = 3.00E8m/s. There are many energy states for electrons in an atom, and thus there are a large number of excited states. When the emitted light of different frequencies is passed through a prism, the frequencies are separated. Each frequency has a definite energy, wavelength, and color (in visible region). The separated light is called a **bright line spectrum**. Each element has it's individual characteristic line spectrum due to the differences in the electronic energy levels among atoms of elements.

A line spectrum, or an atomic spectrum has the following characteristics:

1. The spectrum of each element is unique and unlike that of any other element.
2. The lines in the spectrum of one element are unaffected by the presence of a second element.
3. The intensity (brightness) of the lines is proportional to the quantity of the element present.
4. Only small amounts of an sample (micrograms) are needed to obtain a spectra.

When e- move from various excited states to the ground state, light of different wavelengths are produced. These correspond to different colors of light. Mercury emits a blue line , while sodium emits a distinct yellow line. The equation below relates the wavelength of emitted light to the electronic energy levels of the hydrogen atom. The ground state or lower energy level is designated n_ℓ, and n_h is the upper or higher energy level.

$\nu = R\,(1/n_\ell^2 - 1/n_h^2)$, **Rydberg Equation**

Various versions of the Rydberg equation will be utilized which yield other desired calculations, e.g. $1/\lambda = R\,(1/n_\ell^2 - 1/n_h^2)$ for wavelength calculations.

The Rydberg constant R can have several values. Wavelength can be calculated using an R = 1.097E-2nm^{-1}. One value which can be used to calculate the energy difference when a transition occurs is 2.179E-18J. Another value is 3.289E15s^{-1} for frequency of transition determination. For the hydrogen atom, the n_l value = 2. From the determination of the wavelength of the emitted light, the higher energy states, n_h can be calculated from the Rydberg Equation. ΔE and ν will be calculated from the Rydberg Equation. In this experiment, Planck's Constant will be determined from a plot of ΔE vs ν. The slope of the plot should equal Planck's Constant.

Sample Calculations

$1/\lambda = R(1/n_l^2 - 1/n_h^2)$

$1/\lambda R = 1/n_l^2 - 1/n_h^2$, $n_l = 2$, $1/n_l^2 = 1/2^2 = 1/4 = 0.25$

$1/n_h^2 = 0.25 - 1/\lambda R.$ If λ = 656nm

$1/n_h^2 = 0.25 - 1/\lambda R. = 0.25 - 1/(1.097E\text{-}2 nm^{-1} \times 656 nm) = 0.25 - 0.1386 = 0.1114$

$1/n_h = (0.1114)^{1/2} = 0.334$

n_h = 1/ 0.334 = 2.996, Nearest Whole Number, n_h = 3

Transition: n = 3 to n = 2

ΔE **(Rydberg) = 2.179E-18J** $(1/n_l^2 - 1/n_h^2)$ = 2.179E-18J $(1/2^2 - 1/3^2)$ =**3.03E-19J**

ν = **3.289E15s^{-1}**$(1/n_l^2 - 1/n_h^2)$ = 3.289E15$s^{-1}$$(1/2^2 - 1/3^2)$ = **4.57E14s^{-1}**

The following encomprises the electromagnetic spectrum:

Cosmic rays	Gamma rays	X rays	Ultrav iolet	Visible	Infrared	Microwave	Radio
10^{-5}	10^{-2}	10^{-1}	10^{1}		10^{3}	10^{5}	10^{9} **10^{11}(nm)**

The colored region of the electromagnetic spectrum is the **visible** region. The colors range from approximately 400 nm(Violet) to 700nm(Red).

The experiments will be performed in three parts (A,B, and C).

In Part A,we will observe the characteristic color emitted from several elements and we will use the Spectronic 20 spectrophotometer to match these colors of electromagnetic radiation with their approximate wavelength. Your instructor will tell you how to use the Spectronic 20.

In Part B, Flame tests we will observe the characteristic color emitted from a number of elements. The probability of the electron falling from upper to lower energy levels (de-excitation) is observed and is reflected by the intensity of the line in the spectrum. The predominant color (more intense line) observed is evidence of the greater probability.

In Part C, we will determine the wavelength, λ, E_{photon}, and n_h for each observed line in the Hydrogen atom spectrum, and calculate a value of Planck's Constant.
You must be able to graph data, and to determine the slope of a line.

EXPERIMENTAL PROCEDURE

A. Spectronic 20-Visible Spectrum

Turn on the Spectronic 20 and let it warm up for 10 minutes. Open the cover to the sample compartment and insert a white strip of paper 1.5 cm wide by 11 cm long (cut from a 3.5 x 5 inch file card) until the bottom of the card trips the lever in the bottom of the sample compartment. While turning the wavelength selection knob on the top of the instrument, look directly down into the sample compartment. You should see all the colors of the rainbow as you turn the wavelength selection knob. Starting at 630 nm record each color and the wavelength setting from 630 to 450 nm. Enter the results on your report sheet.

B. Flame Tests

1. Rinse a piece of wire (platinum or nichrome) with a loop on the end with distilled water. Dip the end of the wire with a loop into concentrated HCl. Insert into a bunsen flame in the hottest region and heat until there is no visible color. Repeat this procedure for each test.

2. In a watch glass, add a few crystals of $BaCl_2$ to 5 or 6 drops of conc. HCl and stir. Dip the clean wire into the mixture and place into a bunsen flame-hottest region (tip of inner cone). Observe the color produced and record it . Once completed for all the salts, use the Spectrum Chart to identify or match the wavelengths with the colors observed to obtain the actual line wavelength.

3. Repeat steps 1 and 2 for $CaCl_2$, NaCl, LiCl, and $SrCl_2$ salts.

4. Using a Spectroscope, observe the following light sources and identify the colors seen by looking through the Spectroscope at the Lamps of Mercury and Neon . Note all colors seen.

C. Hydrogen Spectrum

Observe the hydrogen spectrum through a Spectroscope looking at a Hydrogen Lamp. Note all colors seen and record them on the data sheet. Go to the Spectrum Chart and determine the actual wavelengths of the light seen for each color and record it. Complete the table with the colors , wavelengths, ΔE photon , $\mathbf{n_h}$ calculated, $\mathbf{n_h}$ (rounded to the nearest whole number), ΔE and the frequency values will be calculated. A plot of **ΔE (y-axis) vs ν (x-axis)** will be made, the slope of which should equal Planck's Constant.

Safety Precautions: 1. When dipping the wire into the concentrated mixtures, exercise extreme caution .
2. Do not allow the wire to touch your skin. If it touches you, flush with plenty of water.
3. Exercise caution when dropping the concentrated HCl onto the watch glass.

EXPERIMENT 8 SPECTROSCOPY

DATA AND LAB REPORT SHEET

NAME____ASSIGNED #______ DATE________

DATA:

A. SPECTRONIC 20

Colors and Corresponding Wavelengths (nm)

COLORS	WAVELENGTH (nm)
______________	______________
______________	______________
______________	______________
______________	______________
______________	______________

B. FLAME TESTS

SALT	PRINCIPAL COLOR	WAVELENGTH
__________	_____________	__________
__________	_____________	__________
__________	_____________	__________
__________	_____________	__________
__________	_____________	__________

Hg COLORS	_______	________	_______	________	_______	________
WAVELENGTHS	_______	________	_______	________	_______	________
Ne COLORS	_______	________	_______	________	_______	________
WAVELENGTHS	_______	________	_______	________	_______	________

C. HYDROGEN SPECTRUM

Color	λ (nm)	n_l	n_h(nearest whole number	Transition (e.g. n=3-->n=2)	ΔE (use Rydberg) p78	ν (use Rydberg p78)
		2				
		2				
		2				
		2				
		2				

1. Show calculations for any one color for all the items in the part C.

2. Plot ΔE (y-axis-ordinate) vs ν (x-axis -abscissa)

3. Determine the slope from two points (X_1,Y_1) (X_2,Y_2) $m= Y_2-Y_1/X_2-X_1$:
(Show Calculation)

4. Determine **h**, Planck's Constant , **h** = m(slope) (units are J.s)

5. Accepted Value of **h** in J.s = ______________

6. Calculate the % Error =[| E-T| \ T] x 100 =
E= Your Experimental Value, T = Value from Theory(Theoretical Value)=Accepted

Pre-Laboratory Assignment Experiment 8

Spectroscopy Name____ #_____ Score_____

1. Distinguish between absorption and emission.

2. The strong line in the calcium emission line spectrum occurs at 657.3nm.
 a. Calculate the energy of the photons.

b. This transition corresponds to what color?

3. a. Using the transition equation, calculate λ for an electron transition from the $n_h = 2$ to the $n_l = 1$ energy level .

b. In what region of the electromagnetic spectrum does this line appear? ultraviolet, visible or infrared

c. Calculate the photon energy resulting from this transition. ΔE (J / Photon)

4. A large number of hydrogen atoms have their electrons excited to the n=4 energy level. What are all the possible transitions from this level to those below it (n_h to n_l) which yield the spectral lines in the emission spectrum. Encircle those that are in the visible region.

n =4 __________________

n =3__________________

n =2__________________

n =1__________________

5. What does it mean when we say that an atom is in an "excited state"?

6. What is n_l equal to for the hydrogen atom?

10 DIVISIONS PER CENTIMETER

EXPERIMENT 9

MOLECULAR WEIGHT BY VAPOR DENSITY

OBJECTIVES

a) The primary objective is to determine the molecular weight of a volatile compound.
b) The secondary objective of this experiment is to become well acquainted with the ideal gas law.

SUPPLIES AND EQUIPMENT

Bunsen burner, lighter, Unknown volatile liquids, Al Foil, Rubber Bands or Al-wire, Thermometer, 600 ml Beaker, 125 Erlenmeyer Flask, Wire Gauze, Ring stand, Balances

WHAT YOU SHOULD KNOW BEFORE YOU BEGIN

Many molecular compounds with low molecular weights have boiling points below the boiling point of water. A volatile compound is one that vaporizes at low temperatures. A substance which is gaseous at room temperature is said to be a gas. However, if a substance is a liquid at room temperature but has a low boiling point, it can become a gas at above the boiling temperature. This gas is referred as a vapor (i.e., any gaseous substance with a boiling point above room temperature is said to be a vapor).

The properties of gases have been studied extensively, and there are a number of relationships found among them. In particular, it has been found that for an ideal gas the relationships are much simpler. An Ideal or perfect gas is a gas that has no molecular interaction among the molecules. At low pressures (i.e., about atmospheric pressures and below) all gases can be approximated as perfect gases with a minimum error in calculations. Our experiments will be carried out at atmospheric pressure, therefore, we will treat the gas or vapor as perfect.

One of the very useful semi-empirical relationships is the universal gas law: $PV = nRT$ (1)

where (P) represents the pressure of the gas, (V) the volume, (n), the number of moles, and (T), the temperature on the Kelvin scale. (R) in this equation is a constant and its magnitude depends on the units of P and V. Two useful and commonly used values of R are given below:

(R) = 0.08205 L atm/ mol K   (R) = 62.36 L torr/ mol K

It is clear that if you use pressure in atmospheres, the first value of (R) will be used; if torr is used for pressure the second value of (R) will be appropriate. The number of moles of a substance is the number of grams divided by the molecular weight for a molecular compound: $n = g/MW$ (2)

where (g) is the mass of the gas expressed in grams and (MW) is the molecular weight. Combining equations (1) and (2) we obtain:

$$PV = gRT/MW \qquad (3)$$

or

$$MW = gRT/PV \qquad (4)$$

Equation (4) provides us a simple and practical method for determination of the molecular weight of a volatile substance. A volatile compound is a liquid at room temperature. However, some molecules of the compound are in the gas phase. If the compound is placed in a sealed container, then part of the liquid will go into the vapor phase. At any given temperature, the pressure of the vapor over the liquid has a given pressure which is said to be the vapor pressure of that compound at that temperature. By increasing the temperature, more of the liquid will go into the vapor phase. We can continue raising the temperature and observe higher vapor pressures. This will continue up to a temperature so that the meniscus no longer can be seen, which is called the critical temperature of the liquid.

However, the situation is different if the compound is placed in an open container. By increasing the temperature, as before, the vapor pressure will increase, and this increase will continue up to the boiling point of the compound. At the boiling point, the vapor pressure of the liquid is equal to the pressure exerted by the gas above the liquid (which in this case will be equal to the atmospheric pressure. In the container itself, as the temperature is increased, more compound goes to the gas phase thus reducing the air molecules in the container. At the boiling point ,essentially all the air molecules will be driven out of the container. If one continues to boil to the last drop of the liquid, then the container will be completely filled with vapor at that temperature and atmospheric pressure.

Therefore, one can measure the pressure of the gas (barometric pressure), the temperature of the gas (the temperature of the water bath at equilibrium with the vapor), mass of the gas (by cooling the flask and weighing), volume of the flask (by filling the flask with water and measuring the water with a graduated cylinder). Placing these determined values in the equation (4) will yield the molecular weight of the compound.

EXPERIMENTAL PROCEDURE

1. Secure a 600 mL beaker with an iron ring clamped to a ring stand as shown in Figure 9.1.

2. Add about 300 mL tap water and start heating the beaker. The thermometer should be suspended by a copper wire from ring stand support.

3. While the water is warming, weigh a clean 125 mL Erlenmeyer Flask together with a square inch aluminum foil and a rubber band to 0.01g.

4. Add about 3 mL of your unknown in the erlenmeyer flask and cover it with aluminum foil, wrapping the rubber band around the top to secure the aluminum foil tightly. Make a small pin hole in the center of the aluminum foil.

5. When the water gently boils, place the flask inside the beaker with the clamp around the neck and secure it to the ring stand. The flask should be immersed in the water as much as possible without wetting the aluminum foil.

6. When the liquid has completely vaporized, record thermometer reading and remove the flask with the clamp and allow it to cool to room temperature. The flask can be placed under a large beaker for cooling and a thermometer used to ascertain readiness for weighing.

Safety note: Use a low flame when heating and keep the mouth of the flask away from the flame at all times. Flammable vapors are released through the pinhole.

7. Periodically touch the flask with the thermometer and observe if the mercury rises for at least 3 minutes. If the mercury does not move , then and only then should the flask be weighed.

8. Wipe the outside of the flask with a paper towel, and weigh it along with the aluminum foil and the rubber band.

9. Without washing the flask, add 4ml more of the same unknown and repeat the processes for a second time.

10. At the end of the experiment, fill the flask with tap water to the rim and using a graduated cylinder determine the volume of the water which equals the volume of the vapor. Do not measure the volume before the experiment is finished, for it is difficult to dry the flask and re-use it. Convert the volume to liters.

CALCULATIONS

1. Record the atmospheric pressure from the barometer in (inches of Hg) and convert it to (atmosphere, atm) units.
2. The mass of the substance is the difference of two weighings (flask filled with vapor and the empty one).
3. The volume of the flask is determined by the graduated cylinder.
4. The temperature is measured directly and converted to degrees Kelvin.

$$\text{Molecular Weight} = gRT/PV = \frac{\text{Mass of Vapor x 0.08205L-atm/mol K x T(K)}}{\text{Pressure (atm) x Volume (L)}}$$

Safety Precautions: 1. Safety glasses must be worn at all times while in the lab.
2. Hot water is dangerous, the beaker should be secured with a large iron ring to the ring stand.
3. Assume that the compound used is hazardous, do not inhale the vapors.

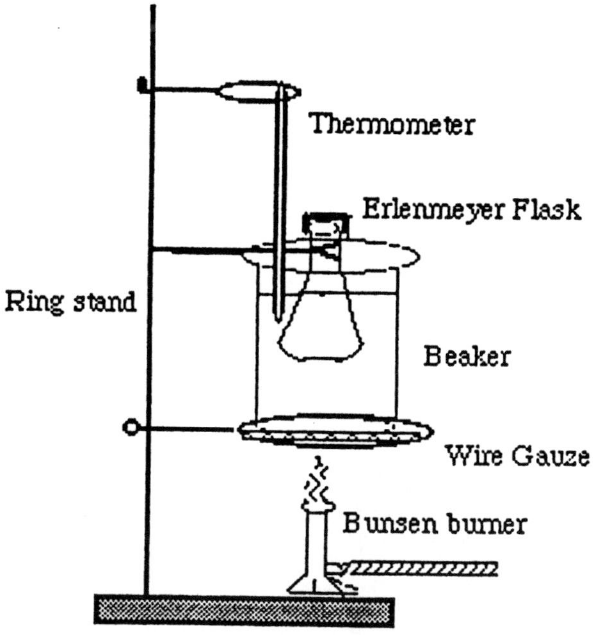

Fig. 9.1 Vapor Density Apparatus

EXPERIMENT 9 MOLECULAR WEIGHT BY VAPOR DENSITY

DATA AND REPORT SHEET

NAME________________________________ASSIGNED #______ DATE__ SCORE_______
UNKNOWN #____________

	Trial 1	Trial 2	
Temperature of Vapor	____________	____________	^{o}C
Temperature of Vapor	____________	____________	^{o}K
Mass of flask + Al + rubber band	____________	____________	g
Mass of flask+Al+ rubber+ vapor	____________	____________	g
Mass of the vapor	____________	____________	g
Barometer reading	____________	____________	inches Hg
Pressure of Vapor	____________	____________	atm

VOLUME MEASUREMENT

	Trial 1	Trial 2	
Volume of Flask	____________	____________	mL
Volume of Flask	____________	____________	L
Molecular Weight of the Unknown	____________	____________	g/mol

Average Molecular Weight __________________g/mol

Theoretical Molecular Weight __________________g/mol
(From Text or Reference)

Percent error _________%

Show Your Calculations Below:

Pre -Laboratory Assignment Experiment 9
Molecular Weight By Vapor Density
Name________________________________#______ Score______

1. In a molecular weight determination experiment, a student measured the volume of the flask to be 146.7 mL, the temperature of the vapor as 98.6°C, the mass of the empty flask 75.42 g and its mass with the condensed vapor was 76.26 g. The barometer reading was recorded as 29.85 in Hg. Calculate the molecular weight of the compound.

2. What would be the pressure of 0.500 g C_2H_6O if all of the compound vaporized in a 512 mL flask at 100°C?

3. In the molecular weight determination experiment
   a. What would be the effect if some air remains in the flask after the compound was completely vaporized?

   b. What can you do to remedy this problem?

   c. Why do we make a small pin hole in the aluminum foil?

EXPERIMENT 10

BOYLE'S LAW

OBJECTIVES

a) To verify if Boyle's Law is applicable to air
b) To determine the pressure and volume of a given amount of a gas (air in this instance) at room temperature
c) To plot volume against the reciprocal of pressure to determine the slope and from that the Boyle's law constant

SUPPLIES AND EQUIPMENT

Buret / group, 00 rubber stopper, 90 cm long Rubber tubing, Thermometer, Meter stick, Plastic ruler, Funnel, Graph Paper

WHAT YOU SHOULD KNOW BEFORE YOU BEGIN

The product of the volume and the pressure of a an ideal gas at any given temperature is constant constant:

$PV = k;$ or

$V = k/P.$ (1)

where P is the pressure, V is the volume of the gas, and k is constant and is equal to nRT. Equation (1) is the equation of a straight line (like y = ax). If we determine the volume and pressure pairs of a sample of gas at various combinations and keeping the temperature constant, a plot of volume, V, versus the inverse of the pressure, 1/P, should yield a straight line. Obtaining a straight line is an indication that the gas is an ideal one.

An easy method of measuring pressures at or near atmospheric pressure or less is by using a U tube. It can be shown that for a liquid or a gas the amount of pressure will be obtained from equation (1).

$P = dgh$ (2)

Where, (d) stands for the density, (g) is the acceleration of gravity ($g = 9.806\ ms^{-2}$) , and (h) is the height of a column of liquid or gas. A barometer which is essentially a 90 cm long test tube is used to measure atmospheric pressures (see Figure 10.1).

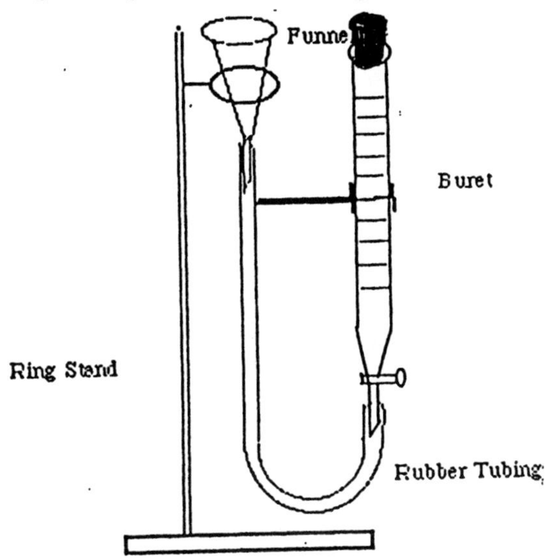

Figure 10.1. Boyle's Law Apparatus

Mercury, a liquid at room temperature has a density of approximately 13.59 g/mL at 0^oC is used in the barometer. A 760 mm high column of mercury exerts a pressure equal to (dgh) and is called 1 atm.

1 atm = 760 mm Hg (3)

But a (mm) is the unit of length. The amount of pressure exerted by 1 mm of mercury is referred to as a torr.

1 atm = 760 torr (4); or 1 mm Hg = 1 torr

A U tube would allow you to measure the pressure of a gas sample. In general, it is called a manometer. An open end manometer is used in measuring the pressure difference between the gas and the atmosphere according to Figure 10.2A.

$$P(g) = P(atm) - dgh \quad (5)$$

where, (h) is the difference in liquid levels. Apparently, the pressure of a gas, in this case, is less than atmospheric pressure. If the pressure of a gas is greater than atmospheric pressure (see Figure 10.2B), equation 6 will hold:

$$P(gas) = P(atm) + dgh - P(H_2O) \quad (6)$$

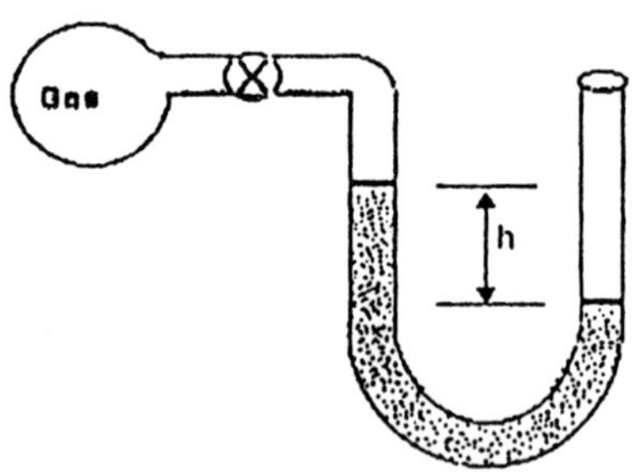

Figure 10.2A. A manometer (Gas Pressure < Atmospheric Pressure)

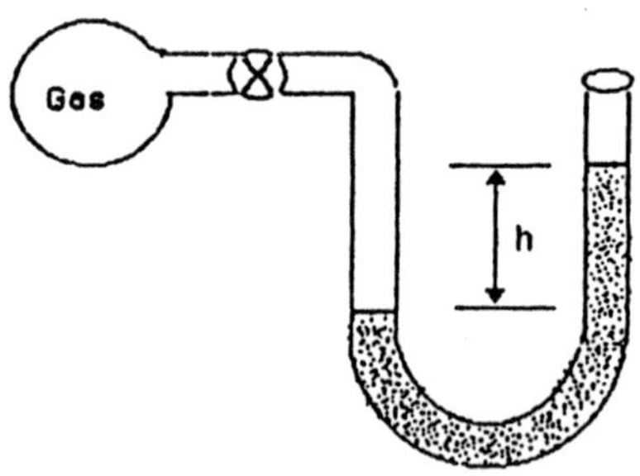

Figure 10.2B. A manometer (Gas Pressure > Atmospheric Pressure)

It is possible to use another liquid instead of mercury. Measuring the difference of the liquid levels in the two arms will yield the pressure difference. To express this difference of height in torr we can use the following equation:

$$(dgh)(water) = (dgh)(Hg) \quad (7)$$

or

$$h(Hg) = (dh)(water)/d(Hg) \quad (8)$$

If we use water, which is plentiful and safe at room temperature with a density of about 1 g/mL, the difference in elevation (for the same pressure difference) will be some 13.59 times larger than that with mercury with a density of about 13.9 g/mL.

EXPERIMENTAL PROCEDURE

Set the apparatus as shown in Figure 10.1. A clean buret (clamped to a ring stand) is connected to a funnel (held by a clamped iron ring) by a one meter long rubber tubing.

1. Lower the funnel down so that it will stand at a lower position than number 50 on the buret.
2. Add distilled water to buret and open the stopcock to let water run to the funnel. At the beginning, you can unclamp the iron ring allowing the water to run down and drive all the air inside the rubber tubing out.
3. At this time clamp the iron ring at about the same height as the number 50 reading of the buret.
4. Add water until the buret reading is about 45.
5. Use a 00 rubber stopper on the top of buret to trap about 50 mL of air. The exact number of milliliters of air is the buret reading plus the correction for the ungraduated portion of the buret on the top of it.
6. Use a meter stick to measure the height of the water level in the buret and in the funnel from the bench top. The height of the funnel minus height of buret is (h).
7. Determine the temperature of the laboratory.
8. Ask your instructor for the barometer reading.
9. Elevate the level of the funnel by about 15 cm and read the difference between the liquid levels. Continue this up to about 75 cm difference in liquid levels.
10. To obtain the correction for your buret, disconnect the rubber tubing and fill it to the rim with water, place the stopper as before and remove it. Place your 10 mL graduated cylinder under the buret and by opening the stopcock adjust the water level to zero. Read the number of mL of water in the graduated cylinder.

SAMPLE CALCULATIONS: 1mm = 1torr

If (h) = 20cm= 200mm, Patm = 29.6inHg, T= $25^{o}C$, d(water)= 0.997= 1g/mL

$$h(Hg) = \frac{d(water) \times h(water)}{d(Hg)} = \frac{1 g/mL \times 200 mm}{13.59 g/mL} = 14.67 mmHg = 14.67 torr$$

$$P(gas) = P(atm) + dgh - P(H_2O) = 29.6 inHg \times \frac{25.4 torr}{1 in} + 14.67 \text{ torr} - 23.8 torr$$

$$P(gas) = 751.8 + 14.67 - 23.8 \text{ torr} = 742.7 torr$$

1. The pressure of the trapped air, P(air), is the barometer reading in torrs plus h(Hg) calculated from equation (6).
2. The gas pressure is P(g) = P(air) + P(water).
3. The volume of the gas is the buret reading + the correction.

Safety Precautions:
1. Safety glasses must be worn at all times during this experiment.
2. Handle all glassware safely and carefully.

EXPERIMENT 10 BOYLE'S LAW

DATA AND LAB REPORT SHEET

NAME________________________________ ASSIGNED #___ DATE_____ SCORE____

Barometer reading_________ inches Hg convert to _________mmHg _________torr

Temperature of the lab _________ ^{o}C _________ ^{o}K

Volume Correction _____________ mL

PRESSURE READING

h(funnel)	h(buret)	h(water)	h(Hg) = h(water)/13.6

VOLUME MEASUREMENTS

Buret reading	Volume of the gas= Buret reading + correction (mL)

PRESSURE VOLUME CALCULATIONS

Pressure(torr)	Volume(mL)	PV(torr mL)	1/P ($torr^{-1}$)

Average of PV = k ____________________

Slope of the Plot of V vs 1/P ____________________

Boyle's Law constant, k ____________________

How does the slope of the V vs 1/P plot relate to PV?

Show Sample Calculations Below:

Pre-Laboratory Assignment Experiment 10
Boyle's Law Name______________________________#__________ Score____________

1. In an experiment investigating Boyle's law, a student used water as the liquid. The buret reading was recorded as 22.62mL, and its height was 38.30 cm; the height of water level in the funnel was 59.20 cm. The corrected volume of the top of the buret was determined to be 6.90 mL, and the barometer reading was recorded as 29.89 inches of mercury at 23.2 ^{o}C. Calculate:
   a. The volume of gas in mL.

   b. The atmospheric pressure in torrs.

   c. The volume of gas in liters, L.

   d. The pressure of dry gas in torrs.

   e. If the number of moles of gas can be expressed as: $n = PV/(RT)$;
   where, R = 62360 mL torr/ mol K, Calculate the number of moles of dry gas in the buret.

2. An open end manometer containing mercury, was used to measure the pressure of a sample of hydrogen gas which was collected in an experiment. The mercury level in the open end of the manometer was 3.76 cm lower than the gas end. If the barometer reading was 29.60 in Hg, calculate the gas pressure in torrs.

3. What is the difference between a barometer and a manometer?

10 DIVISIONS PER CENTIMETER

EXPERIMENT 11

PAPER CHROMATOGRAPHY

OBJECTIVES

a) To become familiar with the techniques of chromatographic separation of food color dyes and/or metal cations using filter paper or other chemical substances
b) To perform R_f calculations

SUPPLIES AND EQUIPMENT

Ascending Paper Chromatography:
Food Coloring, Test Tubes-3 (1.3cm x 15cm), Solid Rubber Stoppers-2 (Size 0), Filter Paper Strips-4 (1.2cm x 13.5cm), Capillary Pipettes-2, Solvent (90% acetone, 10% 6MHCl), 1% Dimethylglyoxime in ethanol , Required Cations[0.4M $Fe(NO_3)_3$, 0.4M $Ni(NO_3)_2$, 0.4M $Cu(NO_3)_2$]
Circular Paper Chromatography
Falcon Petri Dish-1, Whatman Filter Paper-1 (9cm Diameter), Capillary Pipette-2, 0.1% NaCl

WHAT YOU SHOULD KNOW BEFORE YOU BEGIN

Chromatography is a separation technique that depends upon the differential distribution of the components in a given mixture between a mobile phase and a stationary phase. If the stationary phase is a packed column, it is called column chromatography. If it is a thin-layer attached to a solid backing it is called thin-layer chromatography. If the stationary phase is a filter paper, then it is called paper chromatography. The mobile phase can be a liquid substance (liquid chromatography), or it can be an inert gas (gas chromatography). All chemists, biologists, nutritionists, nurses and engineers should have a fundamental knowledge of chromatography because they must be concerned with the identity and purity of substances they have to work with.

In this experiment, we will be performing paper chromatography ; (A) Ascending Paper Chromatography and (B) Circular Paper Chromatography. Filter paper contains a cellulose structure which must be wet with the solvent-system that is used in the separation so as to allow the mobile phase to move along the paper fibers by capillary action. The **solute is spotted** with a capillary pipette at the bottom of the paper strip, and marked with a pencil or other insoluble substance. A **solute** is the species that is being dissolved. A **solvent** is the dissolving medium. The solute in the **mobile phase** moves with the solvent while the chromatographic development occurs. While the solute is moving, it is experiencing several distributions between the mobile and the solvent phases. The time that the solute remains in the mobile phase determines the rate at which it travels. As the solute moves, if it remains in the solvent phase then it will travel with the solvent front as is illustrated below in Figure 11.1. If it remains mostly in the stationary phase, then the distance travelled by the solute will be very near the initial spotting. The solvent front is noted once the movement has stopped. When the developing process has finished, the solute/distance/s are noted and recorded. A ratio is computed by the following formula:
R_f = distance travelled by the solute / distance travelled by the solvent (front). If the spots are colored, it should be relatively easy to locate them. Where the solutes are colorless, several methods may be used to locate them. One is to spray the strip with a reagent which reacts with the solutes. Another involves the use of a reagent such as iodine vapor which causes organic substances to become visible. Another method is to place the strip under an ultraviolet lamp where many substances can be located as a result of fluorescence.

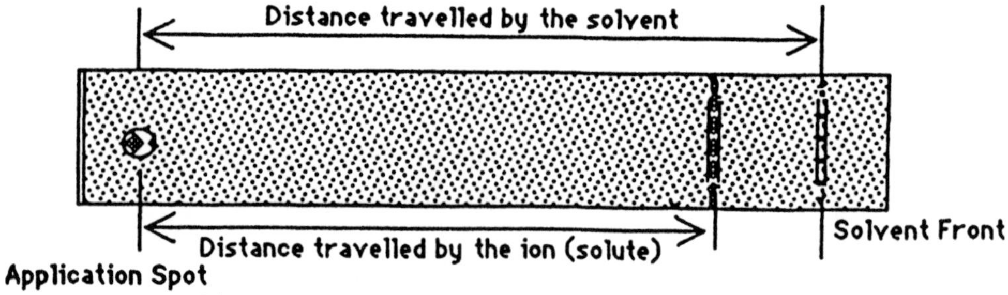

Fig. 11.1 Ascending Strip

Three metal cations will be analyzed in part A of this experiment, Cu^{+2}, Ni^{+2}, and Fe^{+3}. Cu^{+2} is light blue in aqueous medium and when complexed with ammonia (NH_3) is dark blue. A complex ion is produced: $Cu(NH_3)_4^{+2}$. Ni^{+2} (light green) reacts with dimethylglyoxime (DMG), to form a cherry red color species. A complex compound is formed:

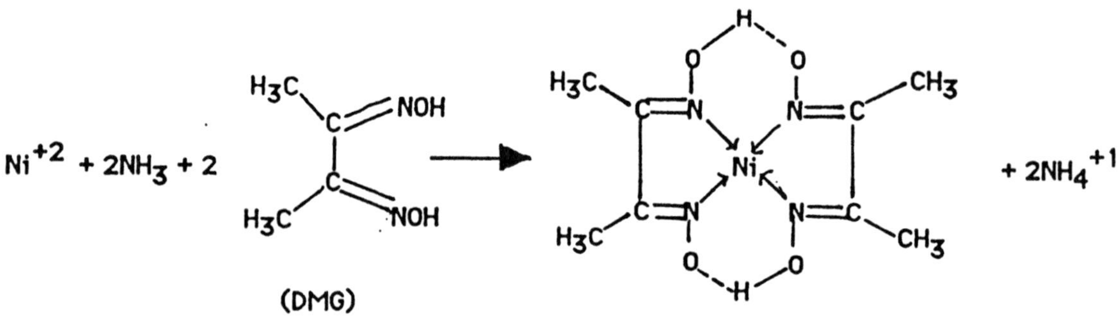

Fe^{+3} in aqueous medium is of a slight yellow color. When combined with the hydroxide ion it forms a reddish-rust compound ($Fe(OH)_3$).

EXPERIMENTAL PROCEDURE

A. ASCENDING PAPER CHROMATOGRAPHY

1. Obtain 4 strips of filter paper approximately 15cm in length. Lightly pencil a 1.5mm diameter circle about 6mm from the bottom of each piece of paper.
2. Obtain a milliliter of each metal ion and the unknown mixture. Dip a capillary pipette in each of the metal ions and spot one strip for each metal ion and one for the unknown by quickly and gently touching the end of the pipette to the center of each circle. The area spotted should not go beyond the circle. It may be necessary to apply more extract. The spot must be dried before each additional application.(Consult Instructor).
3. Finally dry the spot.
4. Pipette 0.5ml of solvent designated by the instructor into the bottom of a test tube as is shown in Fig. 11.2. (Alternatively, an erlenmeyer flask, beaker or other suitable container can be used with appropriate solvent adjustments).
5. Place the spotted paper strip into the test tube. <u>Do not allow it to touch the walls of the test tube.</u> Only the top portion can touch the top of the test tube. The bottom of the paper should dip into the solvent mixture.
6. Stopper the test tube and allow it to remain until the solvent front has traveled to about 5mm from the top of the paper.
7. Remove the stopper and the paper and **mark the solvent front.**
8. Allow the strip to dry.
9. Which metal ion do you detect without further developing? Measure the distance traveled by the solvent front and the distance traveled by that component. Calculate the R_f value.

R_f = distance traveled by solute/ distance traveled by the solvent.

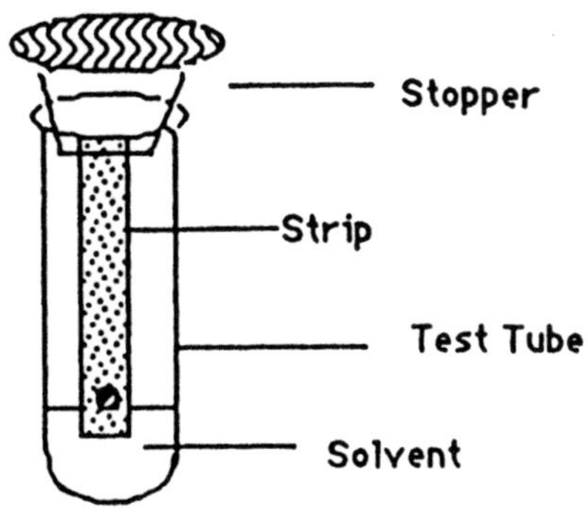

Fig. 11.2 Eluting Chamber

10. The other chromatograms must be developed further under the hood. Pour about 4mL of ammonium hydroxide into a petri dish , and place the filter papers of the other knowns across the dish without allowing any liquid contact. What color develops? Which metal ion is this? Mark the ring front of this ion . Calculate the R_f value.
11. Dip a piece if filter paper into a small amount of the DMG reagent. Using this as a brush , paint the third known metal chromatogram with this reagent. Alternatively, the DMG reagent can be sprayed onto the chromatogram . What color develops? Which metal ion is this? Calculate the R_f value.
12. Repeat the above sequence with your unknown mixture, note colors, and the distances traveled, and calculate R_f values, and identify the unknown components by comparison to the known results.
13. Attach the chromatograms to the report sheet.

Safety Precautions: 1. No flames are allowed in the laboratory while performing this experiment.
2. Do not allow the solvent to contact your skin

B. CIRCULAR PAPER CHROMATOGRAPHY

1. Lightly pencil a 1.5mm diameter circle in the center of a Whatman No.1 filter paper (9cm) in diameter as is shown in Fig. 11.3.
2. Cut a wedge in the paper to form a strip that extends up to a few mm of the pencil mark.
3. Using a capillary pipette, spot one food color inside the 1.5mm circle described above.
4. Place the solvent (0.1% NaCl) in a Petri dish that is slightly smaller in diameter than the filter paper.
5. Bend back the strip in the filter paper so that it will act as a wick when the filter paper is placed on top of the Petri dish.
6. Immediately cover the paper and hold it in place by the Petri dish cover.
7. When the solvent has reached a point near the edge of the Petri dish, remove the paper and mark the solvent front with a pencil.
8. Determine the **R_f** value of the food color as above.
9. Repeat the above procedure with two other food colors.

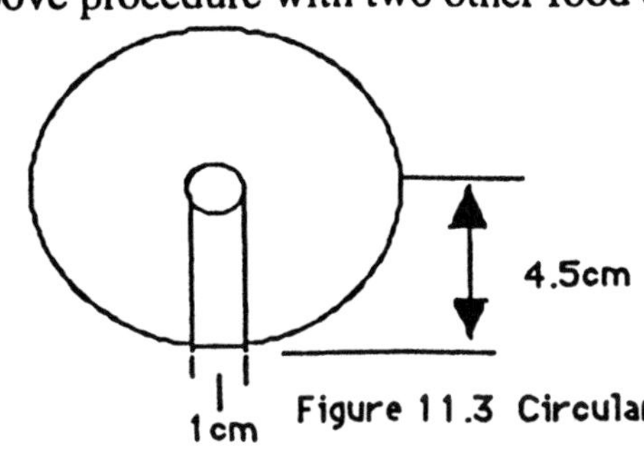

Figure 11.3 Circular Chromatogram

EXPERIMENT 11 PAPER CHROMATOGRAPHY

DATA AND LAB REPORT SHEET

NAME_______________________________ ASSIGNED #______ DATE______ SCORE______

A. ASCENDING PAPER CHROMATOGRAPHY

Spots	Fe^{+3}	Cu^{+2}	Ni^{+2}	Unknown	Unknown	Unknown
Color						
Distance travelled by metal ion						
Distance travelled by solvent						
Rf Values						

Show Calculations for one component:
Give the developing reaction/s for each ion that required developing, and name the reactants and products on lineless paper.

B. Circular Paper Chromatography

Spots	1	2	3
Color	________	________	________
Distance Travelled Solute	________	________	________
Distance Travelled Solvent	________		
Rf Values	________	________	________

1. Is there a pure green food color?

2. Why is a pencil used to mark the points of application?

Pre-Laboratory Assignment Experiment 11
Paper Chromatography Name_________________________________#_____ Score___

1. How many milliliters (ml) of acetone are needed to prepare 20ml of (90% acetone, 10% 6M HCl) by volume?

2. What is paper chromatography?

3. What is meant by the solute and the solvent in paper chromatography?

4. In part A , what would happen to the identification of the substances in the metal cation mixture if the paper is spotted, placed in the solvent , and the spot ends up below the solvent level?

5. Write the equation for the formation of the complex ion, $Cu(NH_3)_4^{+2}$ in aqueous medium?

EXPERIMENT 12

MOLECULAR WEIGHT BY FREEZING POINT DEPRESSION

OBJECTIVES

a) To determine the molecular weight of a nonvolatile, nonelectrolyte
b) To observe the freezing point of a pure solvent
c) To determine the freezing point of a solution
d) To determine the value of ΔT_f (difference between the freezing points of a solvent and a solution)

SUPPLIES AND EQUIPMENT

1-400-ml Beaker, 1-600-ml Beaker, 1-150-mm Test Tube (Hard Glass), Wire Stirrer, Thermometer, Two-hole Stopper, 1buret clamp, Stand, Ice, Water, Cyclohexane, Unknown , Graph Paper

WHAT YOU SHOULD KNOW BEFORE YOUR BEGIN

The addition of a nonvolatile nonelectrolyte solute to a solvent produces characteristic changes in a solvent's physical properties. We know that antifreeze (ethylene glycol) added to an automobile's cooling system prevents freeze-up in the winter and boiling over in the summer because the antifreeze-water solution has a lower freezing point and a higher boiling point than pure water. A plot of vapor pressure as a function of temperature is indicated below. The vapor pressure of a liquid is the pressure exerted by the vapor at equilibrium. The solid line refers to that of pure water, the dashed line to an aqueous solution.

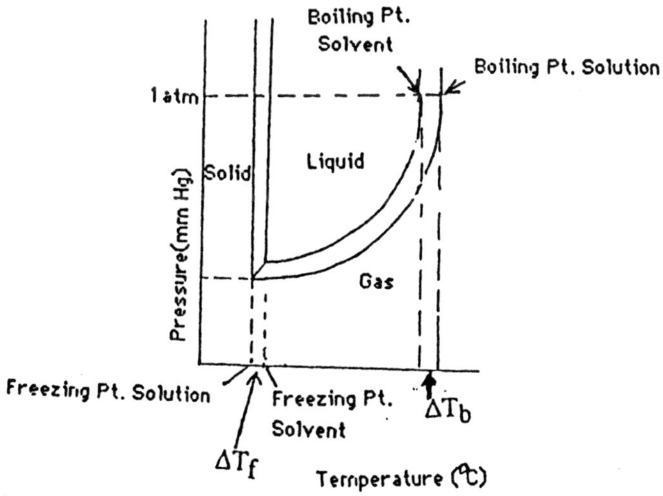

Fig. 12.1 Solvent and Solution Phase Diagram

Pure water boils at 100°C when its vapor pressure equals 760 Torr or 1 atmosphere. When a nonvolatile solute is added to water forming a solution, solute molecules occupy a part of its surface area. This inhibits movement of some water molecules into the vapor state causing a vapor pressure lowering of the water (see figure above), lower than 760 Torr. Since the solution's vapor pressure is less than 760 Torr, it does not boil at 100°C.

For a solution to boil at 1 atmosphere pressure, its vapor pressure must be increased to 760 Torr; this is done by increasing the solution's temperature above 100°C. This is the boiling point elevation of the water due to solute.

The freezing point of water is depressed when a nonvolatile solute is added to it. Water's freezing point is 0°C. The freezing point depression is the difference between the freezing point of pure water and that of the aqueous solution as a result of the addition of the solute.
Vapor pressure lowering, boiling point elevation, and freezing point depression are colligative properties of solutions. These properties are based on the number, rather than the type, of solute particles dissolved in the solvent.

The freezing point lowering, ΔT_f, resulting from the solute addition to the solvent is proportional to the molality, **m**, of the solute:

$$\Delta T_f = | t_f\text{(solvent)} - t_f\text{(solution)} | = K_f m \qquad (1)$$

$$\Delta T_b = | t_b\text{(solution)} - t_b\text{(solvent)} | = K_b m$$

$$m = \frac{\text{mol solute}}{\text{kg solvent}} = \frac{\text{(g/Mol. wt.) solute}}{\text{kg solvent}} \qquad (2)$$

$$\text{Molecular weight, } M = \frac{K_f \times g\text{(wt. of solute)} \times 1000}{\Delta T_f \times G\text{(wt. of solvent)}} \qquad (3)$$

K_f is the molal freezing point depression constant for the solvent. See Table 1 below:

Table 12.1. Molal Freezing Point Depression Constants , Freezing Points, Molal Boiling Point Constants, and Boiling Points of Several Solvents

Substance	Freezing Point °C	Freezing Point Constant, K_f (°C.kg/mol) =°C/m	Boiling Point °C	Boiling Point Constant, K_b (°C.kg/mol)=°C/m
naphthalene	80.2	6.9	218	____
Camphor	179	39.7	207.42	5.61
p-Dichlorobenzene	53.1	7.10	____	____
Cyclohexane	____	20.0	80.7	2.79
Acetic acid	16.6	3.90	118.2	3.07
Water	0.0	1.86	100.0	0.51

In this experiment, you will determine the molecular weight of a nonvolatile solute in cyclohexane by measuring the freezing point for the pure cyclohexane and for the solution. You will also determine the mass of solute and solvent. Using K_f, you will calculate the molecular weight from equation (3).

Cyclohexane's freezing point and that for the solution are obtained from Temperature °C vs Time plots. The cooling curve of cyclohexane reaches a plateau at its freezing point: extrapolation of the plateau to the temperature axis determines its freezing point. The freezing points for the solution and for pure cyclohexane are determined at the intersection of the two lines drawn tangent to the curves above and below the freezing point as shown in Figures 2 and 3 below. The frezing point is sometimes difficult to determine visually when supercooling occurs. Supercooling is when a substance while cooling goes below it's freezing point without solidification or crystallization occurs. Supercooling should and can be minimized with constant stirring.

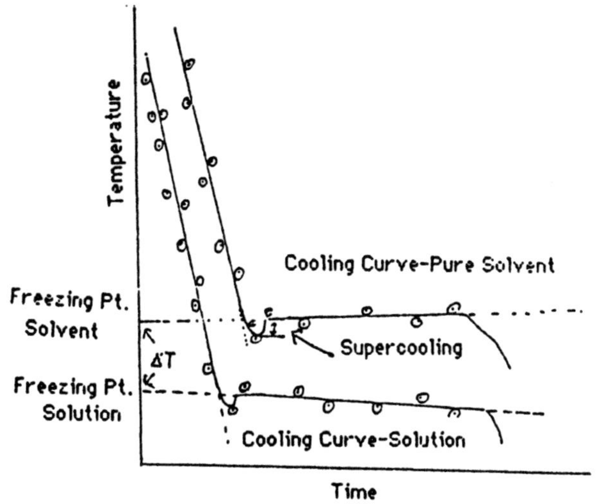

Fig. 12.2 Cooling Curves with Supercooling

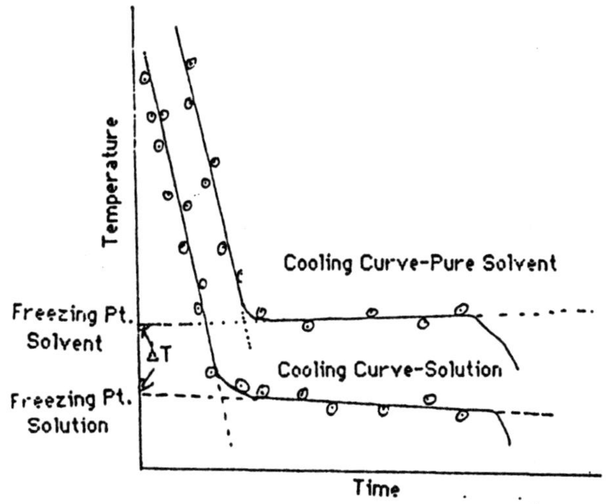

Fig. 12.3 Cooling Curves without Supercooling

EXPERIMENTAL PROCEDURE

A. FREEZING POINT OF CYCLOHEXANE

1. Weigh a clean, dry 150-mm test tube in a 250-ml beaker as shown in Figure 4 and record it.
2. To the test tube, add 15g of cyclohexane and reweigh the tube and contents ($\pm$0.01g) and record it.
3. Assemble the apparatus shown in Figure 5 below.
4. Add to a 400-ml beaker 300-ml of an ice water slurry.
5. Insert a thermometer through one hole of a two hole stopper so that the thermometer would be placed into the solvent to measure the temperature.
6. Insert a wire stirrer through the other hole.
7. Place the test tube with cyclohexane in the ice-water bath quickly , read the thermometer, and record the the temperature at time zero as shown in Fig. 3.
8. While stirring with the wire stirrer inserted, record the temperature readings at time intervals (30 sec or 60 sec) on the report sheet. The temperature will remain constant at the freezing point until the cyclohexane freezes or turns into a solid.
9. Continue recording until the temperature begins to drop again.

10.Remove the test tube out of the ice water bath and allow it to warm up (that was your practice run). Now repeat it for real!

11. On graph paper, plot the temperature ,^{0}C, on the (vertical axis) verses time on the (horizontal axis) to obtain cyclohexane's "cooling curve." The temperature can be taken from the plateau of the cooling curve which is cyclohexane's freezing point. (Get instructor's approval of graph.)

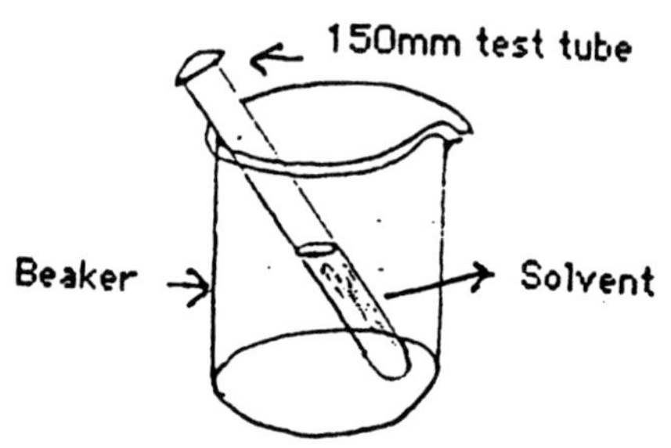

Fig. 12.4 Solvent Weighing Apparatus

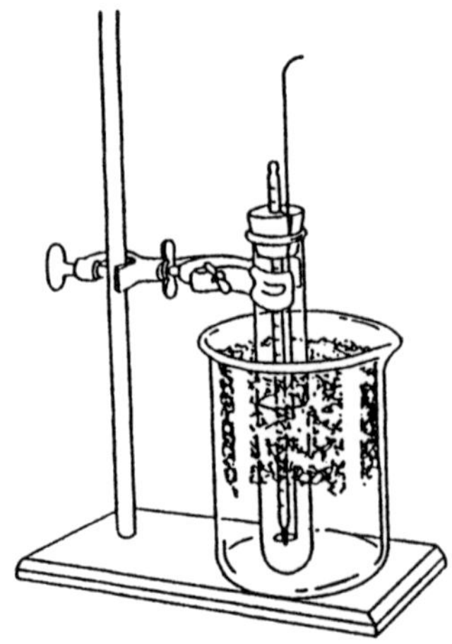

Fig. 12.5 Freezing Point Assembly

B. FREEZING POINT OF CYCLOHEXANE + UNKNOWN SOLUTE

1. Get the solid unknown sample from the instructor. Weigh between 0.1g to 0.3g of the unknown on weighing paper.

2. Transfer the unknown quantitatively to the cyclohexane in the 150-mm test tube shown in Figure 4. Be certain that none of the unknown compound adheres to the test tube's side. If this happens, roll the test tube until the unknown dissolves. If the unknown is a liquid, weigh about 2-ml of it in a 10-ml graduated cylinder and pour the unknown into the cyclohexane. Reweigh the empty cylinder. The difference is the mass of the unknown liquid added to the cyclohexane. (The weight of the unknown liquid can also simply be determined by taring the balance).

3. The freezing point of the solution is determined the same way as the solvent's was found in Part A. When the solution is near the freezing point, record the temperature at intervals (15 to 20 sec).

4. Plot the temperature vs time (on the same graph as in Part A) to get the solution's cooling curve. It will show a "break" at the temperature where freezing begins. Draw lines tangent to the curves above and below the freezing point; the intersection of the lines is the solution's freezing point.

5. Repeat the procedure for Trial 2 by adding to the solute-cyclohexane solution an additional 0.1 to 0.2g of unknown solute as described in Trial 1. If your unknown is a liquid, repeat the procedure described for its measurement and transfer it according to Trial 1. Repeat the freezing point determination and plot the data on the same graph paper using a different color. Precaution: 1. Discard the solution in a waste container that will be provided for you.

COOLING CURVE DATA

SOLVENT				SOLUTION			
<u>PRACTICE 1</u>		TRIAL 1		<u>TRIAL 1</u>		<u>TRIAL 2</u>	
TIME	TEMP	TIME	TEMP	TIME	TEMP	TIME	TEMP

10 DIVISIONS PER CENTIMETER

EXPERIMENT 12 MOLECULAR WEIGHT BY FREEZING POINT DEPRESSION

DATA AND LAB REPORT SHEET

NAME____________________________ASSIGNED #_______DATE________SCORE______
UNKNOWN #______

A.FREEZING POINT OF CYCLOHEXANE

1. Weight of test tube + cyclohexane (g) _______
2. Weight of test tube (g) _______
3. Weight of cyclohexane (g) _______
4. Freezing point from cooling curve (0^oC) _______
5. Instructor's Approval of Graph _______

B. FREEZING POINT OF CYCLOHEXANE + UNKNOWN

	Trial 1	Trial 2
1.Weight of solute (g)	______	______
2. Freezing point from cooling curve (0^oC)	______	______

C. CALCULATIONS

	Trial 1	Trial 2
1. Molal freezing point constant, K_f, of cyclohexane	$20.0^oC\cdot$ kg/mol	
2. Freezing point change, ΔT_f (0^oC)	______	______
3. Weight of solute in solution	______	______
4. Weight of cyclohexane in solution (g)	______	______
5. Molecular weight (g/mol) -Equation (3)	______	______
6. Average Molecular weight (g/mol)	_______________	

QUESTIONS

1. If the weight of solute was less than the correct value, what effect if any would this have on the Molecular Weight ? Why?

2. If the change in the freezing point, ΔT_f was greater than the correct value, what effect if any would this have on the Molecular Weight ?

Prelaboratory Assignment Experiment 12
Molecular Weight by Freezing Point Depression Name______________________#___ Score___

1. A 0.573g sample of a nonvolatile solute is added to 15.6g of naphthalene and the mixture warmed until melted. The freezing point of the solution was determineds and found equal to 78.1°C. The freezing point of naphthalene is 80.2°C. What is the molecular weight of the solute?

2. a) The boiling pt. of the solution in question 1 will ____________________
increase/decrease
b) The vapor pressure of the solution in question 1 will ____________________
increase/decrease

3. What is meant by the term "colligative property"? List three properties

______________ ______________ ______________

4. A. What is the new freezing point of a 0.5m aqueous solution of $C_6H_{12}O_6$?
$\Delta T_f = T_f(\text{normal}) - T_f(\text{depressed}) = K_f \times m$

B. What is the new boiling point of a 0.5m aqueous solution of $C_6H_{12}O_6$?
$\Delta T_b = T_b(\text{elevated}) - T_b(\text{normal}) = K_b \times m$

5. Define undercooling (supercooling).

6. Arrange the following aqueous solutions in order from highest to lowest freezing point. Why?

0.1m NaCl, 0.1m $C_{12}H_{22}O_{11}$, 0.1m $Ba(NO_3)_2$

EXPERIMENT 13

ACID-BASE TITRATION-EQUIVALENT WEIGHT OF AN ACID

OBJECTIVES

a) To prepare and accurately determine the normality of a sodium hydroxide solution
b) To determine the equivalent weight of a solid acid

SUPPLIES AND EQUIPMENT

50-ml Buret, two 250-ml Erlenmeyer flasks, 400-ml Beaker, 1Sheet of white paper, Funnel, Potassium hydrogen phthalate, $KHC_8H_4O_4$, 50% NaOH solution, Phenolphthalein solution, Balances, Unknowns

WHAT YOU SHOULD KNOW BEFORE YOU BEGIN

One of the most important techniques for chemical analysis is titration to an equivalence point. The **equivalence point** occurs when the equivalents of H^+ exactly equal the equivalents of OH^-. When neutralization is carried out in such a manner that the exact volumes of both acid and base used are measured, the process is called titration. In order to know experimentally when the endpoint has been reached, an indicator is utilized or a pH meter. The **end point** in a titration occurs when the indicator changes color. Theoretically the end point should occur simultaneously with the equivalence point ; when the acid equivalents just equal the equivalents of base. The normality of solution is defined as the number of equivalents / liter of solution.

A widely used procedure for preparing and standardizing solutions for titrations is to prepare a solution of approximately the desired concentration and then to titrate it against an accurately weighed quantity of a compound of known purity. The compound of known purity used in this manner is called a primary standard. Nearly all standard solutions of acids and bases are prepared by this method.

In this experiment the end point of the acid-base titration is detected using phenolphthalein indicator, which is colorless in acidic and red (or pink) in basic solutions. Indicators are selected so that the reaction's equivalence point and the indicator's endpoint occur at essentially the same pH in the titration.

STANDARDIZATION OF NaOH SOLUTION

Dry potassium hydrogen phthalate, $KHC_8H_4O_4$, is used as the primary acid standard for determining the normality of a sodium hydroxide solution. This standard is used because of its high purity. The accurately-weighed acid sample is dissolved in distilled water. The equivalents of $KHC_8H_4O_4$ used for the titration is calculated from the acid's weight and its equivalent weight:

$$\frac{\text{mass of } KHC_8H_4O_4 \text{ (g)}}{\text{equivalent weight of } KHC_8H_4O_4 \text{ (g/eq)}} = \text{eq } KHC_8H_4O_4$$

The equivalent weight of $KHC_8H_4O_4$ is 204.2 g/eq. The NaOH solution is then added from a buret to the $KHC_8H_4O_4$ solution until the end point is reached. Your instructor will demonstrate the care and operation of a buret in a typical titration. This is indicated by the phenolphthalein indicator changing from colorless to a barely perceptible pink color. The volume of NaOH solution added to the $KHC_8H_4O_4$ solution is recorded at this point. The normality of the NaOH solution is calculated from the following equation because one equivalent of $KHC_8H_4O_4$ reacts with one equivalent of NaOH:

$$KHC_8H_4O_4 + NaOH \text{ --------> } H_2O + KNaC_8H_4O_4$$

equivalents of $KHC_8H_4O_4$ = equivalents of NaOH

$$\text{Normality of NaOH} = \frac{\text{equivalents of NaOH}}{\text{liters of solution}}$$

EQUIVALENT WEIGHT OF AN ACID

The equivalent weight of an unknown acid is also determined. The standardized NaOH solution is used to titrate an accurately-measured amount of the unknown acid to the equivalence point. The equivalents of acid in the unknown are determined by knowing the volume and the normality of the NaOH. The following equations are used:

$$\text{Volume (liters)} \times \text{Normality} \frac{\text{(equivalents)}}{\text{liter}} = \text{equivalents of NaOH}$$

equivalents of NaOH = equivalents of Acid

From the measured mass of the unknown acid titrated, the equivalent weight of the acid is calculated in the following manner:

$$\text{Equivalent wt. of Acid} = \frac{\text{mass of Acid (g)}}{\text{equivalents of Acid}}$$

EXPERIMENTAL PROCEDURE
A. STANDARDIZATION OF NaOH

1. To prepare 500 ml of approximately 0.1N NaOH, add 2-2.5ml of 50% NaOH to 500 ml of boiled cooled distilled water, and store it in a plastic container. The water should be practically CO_2 free. The CO_2 free water is prepared by boiling about 500-ml of distilled water in a clean 600-ml beaker for approximately five minutes. The water is cooled by placing it under running tap water, taking care not to shake the water unnecessarily, which would permit carbon dioxide to dissolve in it again. (The 50% sodium hydroxide will be prepared by the instructor).
The diluted sodium hydroxide must now be standardized; its exact concentration must be determined. For this purpose, we titrate against a known weight of potassium hydrogen phthalate.

   Precaution: The 50% NaOH is very caustic. Avoid skin contact. If any gets on your hands, wash them thoroughly under water immediately.

2. Obtain a dried sample of pure potassium hydrogen phthalate, $KHC_8H_4O_4$, from the instructor.

3. Prepare two clean 250-ml Erlenmeyer flasks and weigh two $KHC_8H_4O_4$ samples while occupying the balance. Weigh 0.4g to 0.5g of $KHC_8H_4O_4$ and add each sample to each flask. Record all weighings.

4. Add 50-ml of previously-boiled, distilled water and 2 drops of phenolphthalein to each flask. Be sure to mark the flasks so that you will know which flask contains which sample.

5. Wash a 50-ml buret thoroughly with soap and water using a long buret brush. Also, thoroughly clean a funnel. Flush the buret with tap water and rinse several times with distilled water. A clean buret shows no water droplets clinging to its inner wall. Rinse the buret with three 5-ml portions of the NaOH solution, making certain that the NaOH wets its entire inner surface. Drain the NaOH through the buret tip. Get the instructor's approval of your apparatus before continuing.

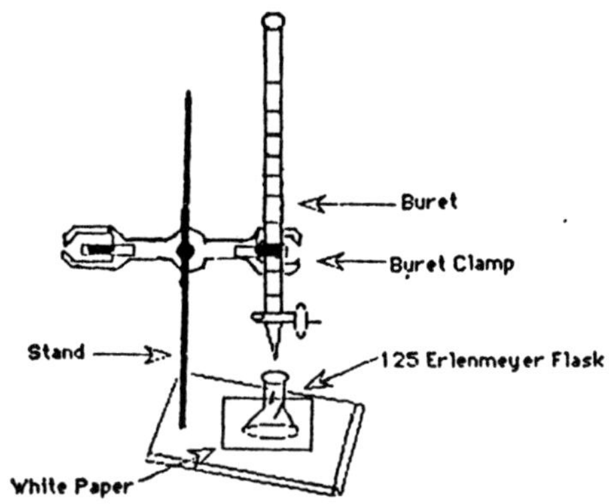

Fig. 13.1 Titration Apparatus

6. Fill the buret with the NaOH solution using a funnel. (Be certain all air bubbles are removed from the buret tip.) and record the initial buret reading. Do this by holding a card with a black line in back of the buret and reading the bottom of the meniscus.

7. Place one of the flasks containing the dissolved potassium hydrogen phthalate, $KHC_8H_4O_4$, under the buret, and place a sheet of white paper under the flask to provide a suitable background for observing the color change while the base solution is being added.

8. Read the **initial meniscus bottom** from the buret and record it as the **initial buret reading.**

9. Slowly add drops of NaOH solution to the acid, swirling the flask after each drop. Mix the solution in the flask by giving the flask an even, circular motion. Consult the instructor for the proper techniques for titrating with the left hand (if right-handed), and adding half-drops from the buret when the endpoint is near. The base solution may be added in a rapid stream of drops until it is noted that the pink color begins to linger while the solution is swirling in the flask. At this stage the solution should be added so slowly that the color disappears between the addition of each drop and the next. Continue to add base until a barely pink color persists throughout the whole solution for at least 30 seconds. Read the volume of the NaOH solution in the buret and record it as the **final buret reading.**

10. Repeat the titration with the second sample of potassium hydrogen phthalate.

B. EQUIVALENT WEIGHT OF AN UNKNOWN ACID

1. Weigh out two samples of your assigned unknown from 0.25g to 0.35g into two 125 erlenmeyer flasks.

2. Add 50-ml of previously-boiled, distilled water and 2 drops of phenolphthalein to each flask. Be sure to mark the flasks so that you will know which flask contains which sample.

3. Place one of the flasks containing the unknown acid under the buret, and place a sheet of white paper under the flask to provide a suitable background for observing the color change while the base solution is being added.

4. Read the **initial meniscus bottom** from the buret and record it as the **initial buret reading.**

5. Slowly add drops of NaOH solution to the acid, swirling the flask after each drop. Mix the solution in the flask by giving the flask an even, circular motion. Consult the instructor for the proper techniques for titrating with the left hand (if right-handed), and adding half-drops from the buret when the endpoint is near. The base solution may be added in a rapid stream of drops until it is noted that the pink color begins to linger while the solution is swirling in the flask. At this stage the solution should be added so slowly that the color disappears between the addition of each drop and the next. Continue to add base until a barely pink color persists throughout the whole solution for at least 30 seconds. Read the volume of the NaOH solution in the buret and record it as the **final buret reading.**

6. Repeat the titration with the second sample of unknown acid.

Precautions:

1. Although you are using a very dilute NaOH solution, remember that it is a base.
2. Please wash your hands before leaving the laboratory.
3. Please wash the buret with soap and water and rinse thoroughly with tap water, and finally with distilled water before returning it to the storeroom.
4. Please make sure that your instructor examines your titration apparatus, a leaking buret should be replaced.

EXPERIMENT 13 ACID-BASE TITRATION-EQUIVALENT WEIGHT OF AN ACID

DATA AND LAB REPORT SHEET

NAME________________________ASSIGNED # ______DATE_____________SCORE_____

UNKNOWN #______

A. STANDARDIZATION OF NaOH

	Trial 1	Trial 2
1. Mass of Erlenmeyer flask plus $KHC_8H_4O_4$ (g)	______	______
2. Mass of Erlenmeyer flask (g)	______	______
3. Mass of $KHC_8H_4O_4$ (g)	______	______
4. Equivalent weight of $KHC_8H_4O_4$	204.2g/eq	
5. Equivalents of $KHC_8H_4O_4$	______	______
6. Final buret reading of NaOH (ml)	______	______
7. Initial buret reading of NaOH (ml)	______	______
8. Volume of NaOH used (liters)	______	______
9. Equivalents of NaOH	______	______
10. Normality of NaOH (eq/L)	______	______
11. Average Normality of NaOH (eq/L)	____________	

B. EQUIVALENT WEIGHT OF UNKNOWN ACID

	Trial 1	Trial 2
1. Mass of Erlenmeyer flask plus unknown acid (g)	______	______
2. Mass of Erlenmeyer flask (g)	______	______
3. Mass of unknown acid (g)	______	______
4. Final buret reading of NaOH (ml)	______	______
5. Initial buret reading of NaOH (ml)	______	______
6. Volume of NaOH used (liters)	______	______
7. Normality of NaOH(eq/L) **(PART A)**	__________	
8. Equivalents of NaOH used	______	______
9. Equivalents of unknown acid	______	______
10. Equivalent weight of unknown acid (g/eq)	______	______
11. Equivalent weight average of the unknown acid	__________	

Show calculations for one complete trial on lineless paper, very neatly and very nicely.

Prelaboratory Assignment Experiment 13
Acid-Base Titration-Equivalent Weight of an Unknown Acid
Name_________________________________# _______ Score____________

1. Distinguish between an equivalence point and an endpoint.

2. A 0.394g sample of potassium hydrogen phthalate ($KHC_8H_4O_4$) is dissolved in 100-ml of water. If 13.9-mL of sodium hydroxide solution are required to reach the equivalence point, what is the normality, N, of the sodium hydroxide? $KHC_8H_4O_4$'s equivalent weight is 204.2g/equiv.

3. A 0.294g sample of an unknown acid requires 13.4-mL of 0.132N NaOH to reach a phenolphthalein endpoint. What is the unknown acid's equivalent weight?

4. What is the color of phenolphthalein in an acidic solution _______?
   in a basic solution________?

5. What signals the investigator to stop adding the NaOH solution?

6. How would you complete the following word equation?

   Strong Base (aq) + Protonic Acid (aq) -----> Water + _______

EXPERIMENT 14

REDOX REACTIONS

OBJECTIVES

a) To write and balance redox equations by the ion electron method
b) To determine an activity series of metals via redox reactions
c) To study the reactions of metals with hydrochloric acid

SUPPLIES AND EQUIPMENT

Na , Mg ribbon, Copper wire (2cm) , Fe wire, Al wire, Mossy Zn, 0.01M $KMnO_4$, Conc. HNO_3, 0.1M $K_2Cr_2O_7$, 0.1M Na_2SO_3 , 6M HCl, 0.5M I_2/KI, 0.1M $Fe(NH_4)_2(SO_4)_2$ (freshly prepared), 0.1M $Na_2S_2O_3$(freshly prepared), 0.1M $CuSO_4$, 0.1M KI, Bunsen burner, Test tubes, Test tube Holder , Litmus paper, Stirring Rod, 3% Starch

WHAT YOU SHOULD KNOW BEFORE YOU BEGIN

There are numerous reactions which occur that are reduction-oxidation reactions. A **reduction** process is one in which there is a gain of electrons evidenced by a **decrease in the oxidation number,** e.g. (Au^{+3} + $3e^-$ ---> Au). An **oxidation** process is one in which there is a loss of electrons evidenced by an **increase in the oxidation number,** e.g. (Mg----> Mg^{+2} + $2e^-$). According to the Law of Conservation, the electrons lost in one process must be gained in another. The electrons lost must equal the electrons gained. Redox reactions are electron transfer reactions. Wherever there is oxidation, there must be reduction. The individual reactions are called half-reactions. The combination of the two processes is referred to as Redox, and the reaction called an Overall reaction. The following is an example of a redox reaction:

$2MnO_4^{-1}$ + 3 S^{-2} + $4H_2O$-------> $2MnO_2$ + 3S + 8 OH^{-1}

In the equation, the manganese oxidation state changed from (+7) in MnO_4^{-1} to (+4) in MnO_2. This a 3 electron reduction (the oxidation state was lowered). The oxidation state of sulfur changed from (-2) in S^{-2} to (0) in S. This is a 2 electron oxidation (the oxidation state increased).

In the determination of whether an equation is a redox equation, a determination must be made as to whether a change in oxidation number has occurred. The following examples illustrate this important concept.

$AgNO_3$(aq) + HCl(aq)-----> AgCl(s) + HNO_3(aq)

the oxidation states of all species has been determined to be :
+1+5-2 +1 -1 +1 -1 +1+5-2
$AgNO_3$(aq) + HCl(aq)-----> AgCl(s) + HNO_3(aq) In this equation none of the substances underwent a change in oxidation number, thus this equation is not that of a Redox reaction.

$Mg(s) + 2HNO_3(aq)$------> $Mg(NO_3)_2(aq) + H_2(g)$
the oxidation states of all species has been determined to be:

0 +1+5-2 +2 +5-2 0

$Mg(s) + 2HNO_3(aq)$------> $Mg(NO_3)_2(aq) + H_2(g)$ In this equation, there has been a change in oxidation number of several species. Therefore, this equation is that of a **Redox** reaction. In the redox equation, the reactant which causes reduction to occur and is itself oxidized is called the **Reducing Agent,** and the reactant which causes oxidation to occur and is itself reduced is called the **Oxidizing Agent.** Many of the Redox reactions involve color changes, formation of solids, liquids and gases. Many of the redox equations are written as ionic equations. A very common redox reaction is that of the rusting of iron containing substances where iron oxides are formed. Originally oxidation was referred to as the addition of oxygen to a substance, while reduction was referred to as the addition of hydrogen to a substance with the subsequent removal of oxygen, e.g.

$2CO(g) + O_2(g)$---> $2CO_2(g)$ and $WO_3(s) + 3H_2(g)$-----> $3H_2O(g) + W(s)$

REDOX EQUATIONS ARE BALANCED BY THE <u>OXIDATION NUMBER</u> METHOD OR THE <u>ION-ELECTRON</u> METHOD.

The Oxidation Number Method is illustrated using the following example:

$H^{+1}(aq) + NO_3^{-1}(aq) + Cu(s)$--> $Cu^{+2}(aq) + NO(g) + H_2O(l)$

1. Assign the Oxidation State of all species present.

+5-2 0 +2-2 +1 -2

$H^{+1}(aq) + NO_3^{-1}(aq) + Cu(s)$--> $Cu^{+2}(aq) + NO(g) + H_2O(l)$

2. Identify the Oxidation and Reduction electron-transfer processes using arrows indicating the number of electrons transferred and the type process:

--------------------------3e- reduction-------------

+5-2 0 +2-2 +1 -2

$H^{+1}(aq) + NO_3^{-1}(aq) + Cu(s)$-----------------> $Cu^{+2}(aq) + NO(g) + H_2O(l)$

--------2e- oxidation--

3. Determine the Coefficient of the Reduced and Oxidized species:
   Locate the LCM= Least Common Multiple between the electrons transferred
   The LCM = 6
   Coefficient of Cu = 6/2 = **3** Coefficient of NO_3^{-1} = 6/ 3 = **2**

4. Place the Coefficients in front of the substances and Balance the remainder by Trial and Error:

$8H^{+1}(aq) + 2NO_3^{-1}(aq) + 3Cu(s)$-----------------> $3Cu^{+2}(aq) + 2NO(g) + 4H_2O(l)$

The Ion-Electron Method is illustrated with the following examples:

$Ag^{+1}(aq) + Cu(s) \text{---->} Ag(s) + Cu^{+2}(aq)$

1. Identify the Oxidation half reaction and the Reduction Half-reaction:

$Cu(s) \text{------>} Cu^{+2}$ oxidation

$Ag^{+1}(aq) \text{--->} Ag(s)$ reduction

2. Atom Balance the Oxidation half-reaction :

Atom balanced as written

$Cu(s) \text{------>} Cu^{+2}(aq)$

3. Charge Balance the Oxidation half-reaction by determining the total charge of the reactants and that of the products by adding electrons as needed.

$Cu(s) \text{------>} Cu^{+2}(aq) + 2e-$

(0) (+2)

The charge on the left = 0 , while that on the right = +2 therefore the charge on the right must be reduced to (0) by adding 2e- (2 negative charges on the right).

4. Atom Balance the Reduction half-reaction :

Atom balanced as written

$Ag^{+1}(aq) \text{--->} Ag(s)$

5. Charge Balance the Reduction half-reaction by determining the total charge of the reactants and that of the products by adding electrons as needed.

$Ag^{+1}(aq) + 1e- \text{--->} Ag(s)$

(+1) (0)

The charge on the left = +1 , while that on the right = 0 therefore the charge on the left must be reduced to (0) by adding 1e- (1 negative charge on the left).

6. Add the two half-reactions such that the electrons loss in oxidation will equal the electrons gained in reduction:

Multiply the reduction by 2 to maintain : electrons loss = electrons gained

$Cu(s) \text{------>} Cu^{+2}(aq) + 2e-$

$2Ag^{+1}(aq) + 2e- \text{---->} 2Ag(s)$

$Cu(s) + 2Ag^{+1}(aq) \text{------>} Cu^{+2} + 2Ag(s)$

The **Reducing Agent** is **Cu** . The **Oxidizing Agent** is Ag^{+1}.

A summary of the procedure for balancing excess H, and Oxygen, O in various media is given below:

Acidic:

For Oxygen, O, balance each excess oxygen with 1 H_2O to the side needing the oxygen then add H^{+1} to balance the hydrogens on the opposite side.

Basic:

For oxygen, O, balance each excess oxygen by adding 2 OH^{-1} to the side needing the oxygen (e.g. 2 excess O yield 4 OH^{-1}) then add 1/2 the amount of H_2O to the opposite side.

For hydrogen, H^{+1} balance each excess H^{+1} with 1 H_2O to the side needing the H^{+1} and then add 1 OH^{-1} to the other side.

If H^{+1} and H_2O are already written in the equation to be balanced, the equation is understood to be in acidic medium .

If OH^{-1} and H_2O are already written in the equation, the equation is understood to be in basic medium.

Balance the following equation in acidic medium:

$CrO_2^{-1} + ClO^{-1} \text{--->} CrO_4^{-2} + Cl^{-1}$

Atom and Charge balance each half-reaction

$ClO^{-1} + 2H^{+1} + 2e^- \text{--->} Cl^{-1} + H_2O$

$CrO_2^{-1} + 2\,H_2O \text{----->} CrO_4^{-2} + 4H^{+1} + 3e^-$

Multiply : first equation by 3 and the second equation by 2 and add the half reactions

$3ClO^{-1} + 6H^{+1} + 6e^- \text{--->} 3Cl^{-1} + 3H_2O$

$2CrO_2^{-1} + 4H_2O \text{----->} 2CrO_4^{-2} + 8H^{+1} + 6e^-$

$3ClO^{-1} + 2CrO_2^{-1} + H_2O \text{------->} 3Cl^{-1} + 2CrO_4^{-2} + 2H^{+1}$

Balance the following equation in basic medium:

$CN^{-1} + MnO_4^{-1} \text{--->} CNO^{-1} + MnO_2$

Atom and charge balance each half-reaction

$CN^{-1} + 2OH^{-1} \text{--->} CNO^{-1} + H_2O + 2e^{-1}$

$3e^{-1} + 2H_2O + MnO_4^{-1} \text{--->} 4OH^{-1} + MnO_2$

Multiply the first equation by 3 and the second equation by 2 and add the balanced half-reactions

$3CN^{-1} + 6OH^{-1} \text{--->} 3CNO^{-1} + 3H_2O + 6e^{-1}$

$6e^{-1} + 4H_2O + 2MnO_4^{-1} \text{--->} 8OH^{-1} + 2MnO_2$

$3CN^{-1} + 2MnO_4^{-1} + H_2O \text{----->} 3CNO^{-1} + 2MnO_2 + 2OH^{-1}$

EXPERIMENTAL PROCEDURE

A. REDOX REACTIONS

1. Cut and place a small piece of sodium metal into a 50ml beaker containing 20ml of distilled water. Note and record all observations. **(Instructor Demonstration)**

2. Dip a stirring rod into the beaker and touch it onto a piece of red litmus paper on a watch glass, and note any changes. Write a balanced equation for the reaction. What does the litmus test reveal?

3. Obtain a polished piece of magnesium ribbon (1cm) and using the tongs insert it into a bunsen flame. Do not look directly at the flame. Note observations. Write a balanced equation for the reaction of magnesium with oxygen. Identify the oxidizing and the reducing agents.

B. REDOX REACTIONS: IONIC EQUATIONS

As you conduct the experiment, please note the colors and physical states of various species given on the following page.

1. Add 20 drops of 0.01M $KMnO_4$ into a test tube, and pour 20 drops of 0.2M Na_2SO_3 into a second one. Using a dropper, add the Na_2SO_3 to the $KMnO_4$ until the permanganate $KMnO_4$ discolors. Wait approximately five minutes and note all observations. Write the balanced ionic equation in basic medium.

2. Add 20 drops of 0.01M $KMnO_4$ into a test tube, and pour 20 drops of 0.2M Na_2SO_3 into a second one. Pour 2mL of 6M HCl into a test tube. Using a dropper, add 5 drops of HCl to the $KMnO_4$, and then add several drops of Na_2SO_3. Note and record all observations. Write the balanced ionic equation in acidic medium. What is the oxidation number of Mn in MnO_4^{-1} and that of S in SO_3^{-2} ?

3. Add 20 drops of 0.01M $K_2Cr_2O_7$ into a test tube, and pour 20 drops of 0.2M Na_2SO_3 into a second one. Pour 20 drops of 6M HCl into a test tube. Using a dropper, add 3 drops of HCl to the $K_2Cr_2O_7$, and then add all of the Na_2SO_3. Note and record all observations. Write the balanced ionic equation in acidic medium. What is the oxidation state of Cr in $Cr_2O_7^{-2}$?

4. Add 20 drops of 0.5M I_2/KI into a test tube, and add 4mL of 0.1M $Na_2S_2O_3$ into a second one. Using a dropper, add the 0.1M $Na_2S_2O_3$ to the I_2/KI until the iodine is discolored. Note and record all observations. Write the balanced ionic equation. What is the oxidation state of S in $S_2O_3^{-2}$?

5. Add 20 drops of concentrated HNO_3 into a test tube. **(Instructor Demonstrated)** Add a piece of copper metal (1/2 cm) into the test tube, and observe for a few moments. Note all observations. The gaseous NO_2 gas will be formed. Conduct under the hood and with proper ventilation. Write the balanced ionic equation.

6. Add 10 drops of 0.1M $CuSO_4$ into a test tube, and add 10 drops of 0.1M KI into a second one. Pour the $CuSO_4$ into the 0.1M KI, and note and record all observations. Add 4 drops of 3% Starch and note the result. Heat the test tube contents in a bunsen flame and note the result. Place the test tube under running tap water for a few minutes and note the result. Write the balanced ionic equation.

7. Add 20drops of 0.01M $KMnO_4$ into a test tube, pour 3mL of 0.1M $Fe(NH_4)_2(SO_4)_2$ into a second tube, and pour 4mL of 6M H_2SO_4 into a third tube. Using a dropper, add the 6M H_2SO_4 to the $KMnO_4$, and then add dropwise the $Fe(NH_4)_2(SO_4)_2$ until a change is observed. Write the ionic equation for the reaction.

Safety Precautions:
1. Do not look directly at the burning of magnesium ribbon.
2. The concentrated acids are dangerous Handle the acid containers (dropper bottles) very carefully, do not allow drops to fall on you or the desk top.
3. Properly dispose of reagents as instructor directs.

Colors and States of Some Substances:

Cu	Reddish Metallic Brown(Copper) due to oxide formation
Mg	Silver White Metal
MgO	White
$NO_2(g)$	Reddish Brown Gas
Mn^{+2}	Pinkish to colorless(aq)
MnO_4^{-1}	Purplish (aq)
MnO_2	Muddy Brown-Black Solid
$Cr_2O_7^{-2}$	Orange (aq)
Cr^{+3}	Dark Greenish (aq)
SO_3^{-2}	colorless (aq)
I^{-1}	Colorless
Cu^{+2}	Blue (aq)
$S_2O_3^{-2}$	colorless (aq)
$S_4O_6^{-2}$	colorless(aq)
I_2	Violet-Black Solid
Fe^{+2}	light green to colorless, (aq)
Fe^{+3}	red brown,(aq)
Cu^{+1}	colorless
CrO_4^{-2}	yellow,(aq)

EXPERIMENT 14 REDOX REACTIONS

DATA AND LAB REPORT SHEET

NAME______________________________ ASSIGNED #_____ DATE_____ SCORE_____

A. Redox Reactions

Na Oxidation	Observations	Reduced species	Oxidized species
litmus			

EQUATION:

Mg Oxidation	Observations	Reduced species	Oxidized species

EQUATION:

B. Reactions Ionic Equations

1.	Observations	Reduced species	Oxidized species

EQUATION:

2.	Observations	Reduced species	Oxidized species

EQUATION:

Oxidation State Of Mn _______ and S__________

3.	Observations	Reduced species	Oxidized species

EQUATION:

Oxidation State of Cr___________

4. Observations | Reduced species | Oxidized species

EQUATION:
Oxidation State of S_________

---

5. Observations | Reduced species | Oxidized species

EQUATION:

---

6. Observations | Reduced species | Oxidized species

EQUATION:

---

7. Observations | Reduced species | Oxidized species

EQUATION:

Pre-Laboratory Assignment Experiment 14

Redox Reactions Name____#_____ Score______

1. Balance the following equation by the oxidation number method.

$Fe^{+2}(aq) + MnO_4{}^{-1}(aq) + H^{+1}(aq)$ -------> $Fe^{+3}(aq) + Mn^{+2}(aq) + H_2O(l)$

Balance the following equations by the ion electron method: Underline the reducing agent and encircle the oxidizing agent. Show the Balanced Half-Reactions

2A. $Cl^{-1} + Cr_2O_7{}^{-2}$-------> $Cr^{+3} + Cl_2$ (In acid)

2B. $Bi(OH)_3 + SnO_2{}^{-2}$---> $Bi + SnO_3{}^{-2}$ (In base)

3. In Redox reactions , the reducing agent is ______________ while the oxidizing agent is_______________________.

4. What is the oxidation state of S in K_2SO_4? of N in HNO_2?

5. What is meant by Charge Balancing?

EXPERIMENT 15

ELECTROCHEMISTRY (GALVANIC CELLS)

OBJECTIVES

a) To construct electrochemical cells based on two half-reactions that are joined electrically by a salt-bridge
b) To measure the voltage of the electrochemical cells
c) To show how cell voltages are related to the concentration of the cell components
d) To determine the relationship between temperature and cell voltage
e) To calculate the thermodynamic functions, ΔG, ΔH, and ΔS values for the cell

SUPPLIES AND EQUIPMENT

Voltmeter, Electrodes of Cu, Sn, Pb, Zn, 0.1M $Zn(NO_3)_2$, 0.1M $SnCl_2$, 0.1M $Pb(NO_3)_2$, 0.1M $Cu(NO_3)_2$, 1.0M $Cu(NO_3)_2$, agar, Voltmeter, Hard glass test tubes, Thermometer, 600mL Beaker, Agar, 0.1M KNO_3, Graph Paper, Alligator Clips, 100mL Volumetric Flasks

WHAT YOU SHOULD KNOW BEFORE YOU BEGIN

A chemical reaction involving the transfer of electrons from one substance to another is an oxidation-reduction (redox) reaction. The substance losing electrons is oxidized and the substance gaining electrons is reduced. When a strip of copper metal is dipped into a solution containing silver ions, a black color coats the strip and the colorless original solution begins to turn blue. This is because the silver ions are reduced, producing black silver metal, and the copper metal is being oxidized. The color of the solution becomes blue as blue copper ions replace colorless silver (I) ions.
The individual half reactions are shown below:

$Cu(s)$ ------------> $Cu^{+2}(aq) + 2e^-$ (oxidation) -a node
$2[\ Ag^{+1}(aq) + 1e^-$ -------> $Ag(s)\ ]$ (reduction) - cathode

$Cu(s) + 2Ag^{+1}(aq)$ ---> $Cu^{+2}(aq) + 2Ag\ (s)$ (Overall cell reaction), Equation 1. (1)

The reaction shown in Equation 1 is a redox reaction. Although, the definition of oxidation and reduction implies loss and gain of electrons respectively, no electron flow can be detected in the system described above. To demonstrate the flow of electrons (to be measured by a digital multimeter) we must build an electrochemical cell consisting of a copper electrode (a strip of copper metal), dipping into a solution of copper(II) ions, and a silver electrode (a piece of silver wire) dipping into a solution of silver(I) ions. The two solutions would be connected by a "salt bridge" which will prevent mixing of the two solutions but will allow the flow of ions between the two solutions. The salt bridge maintains electrical neutrality, and thus prevents the build up of charge at any one electrode. Putting a voltmeter (actually a digital multimeter) into this circuit would permit a measurement of the potential (the tendency of electrons to flow from one electrode to the other) of the cell. The flow of electrons produces a current, I which is converted to voltage. The potential difference or voltage is the reading on the meter, and is the driving force of electrons and is referred to as the electromotive force, **emf** of the cell. The electrochemical cell is shown below in Figure 15.1.

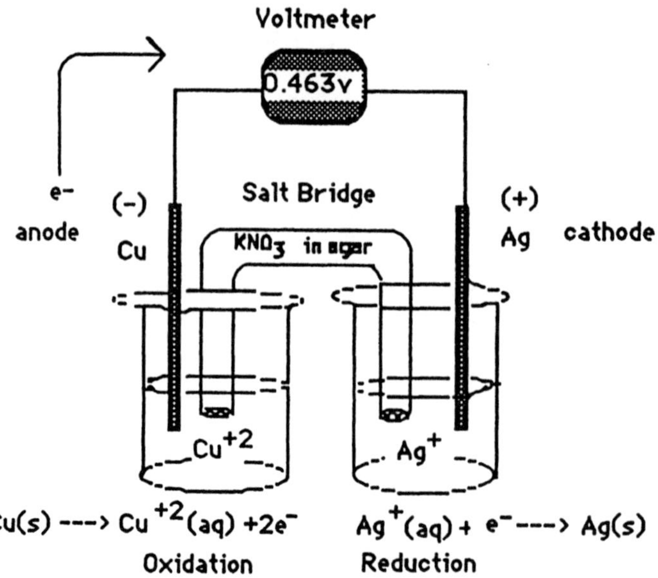

Fig. 15.1 An Electrochemical Cell with a Salt Bridge

The diagram for this cell is : Cu(s) | Cu^{+2}(1M)|| Ag^{+1}(1M) | Ag(s).

Let us look at the oxidation and reduction processes in this cell and find out the origin of the generated electricity. Let us first consider the oxidation half-reaction. In this half-reaction, copper metal is changed to copper ions, each copper atom loses two electrons (see Equation 2).

$$Cu(s) \text{ ---------->} Cu^{+2}(aq) + 2e- \qquad (2)$$

These electrons move through wires, through the voltmeter and around to the silver electrode. The reduction half-reaction occurs simultaneously with the oxidation half-reaction . The silver ions, each gaining an electron, is reduced to form silver atoms (see Equation 3)

$$2[\, Ag^{+1}(aq) + 1e- \text{ ---------} Ag(s)\,] \qquad (3)$$

In the silver half reaction two electrons are transferred to preserve electrical neutrality. The sum of these two half-reactions (see Equations 2 and 3) gives the overall net ionic equation shown in Equation 1.

One can construct similar cells using different electrodes and electrolytic solutions and measure the relative activities of a variety of metals. The voltmeter indicates the cell potential, and also the direction of electron flow. Electrons flow from the anode(-), where oxidation takes place, to the cathode (+), where reduction takes place.The more active metal in the cell forms the anode, while the less active metal forms the cathode. Ions move through the salt bridge, K^{+1} toward the cathode, and the NO_3^{-1} ions toward the anode, to maintain electrical neutrality.

The potential of an electrochemical cell depends upon the nature of the metals used as electrodes and the concentrations of the electrolytic solutions. When the solutions are 1M with respect to all ions in the solution, the cell is designated a "standard cell" and its voltage is called a "standard potential", E^o. The Nernst equation (see Equation 4) must be used to relate the potential of an arbitrary cell to that of the corresponding standard cell.

$$E_{cell} = E^o{}_{cell} - 0.0591/n \log Q \qquad (4)$$

In the Nernst equation, E^o is the standard potential. The symbol E stands for the actual potential (measured by the voltmeter) of a cell in which ion concentrations may differ from one molar. The symbol **n** refers to the number of electrons transferred in the balanced net ionic equation and Q is the reaction quotient, which represents the molarities of the products raised to their coefficient powers divided by the molarities of the reactants raised to their coefficient powers., $[P]^a/[R]^b$. The relation between the potential of any copper/copper salt-silver/silver salt cell to the standard potentials of such cells is given in Equation 5.

$$E_{cell} = E^o_{cell} - 0.0591/2\ (\log [Cu^{+2}]\ /[Ag^{+1}]^2) \qquad (5)$$

In this case the net ionic equation shows that the coefficients "a" and "b" are equal to 1 and 2 respectively. Notice that solids and pure liquids do not appear in the Nernst equation because their concentrations do not change.

The Gibbs Free energy change, ΔG for a reaction yields us a measure of the force that drives a reaction to occur. The degree of spontaniety is indicated by the magnitude and sign of ΔG^o. If ΔG^o is negative (-), the reaction is said to be spontaneous, a positive value(+) indicates nonspontaniety, and a value of zero indicates that the reaction is at equilibrium. The cell potential is related to ΔG^o by equation 6 under standard conditions:

$$\Delta G^o = -\ nFE^o \qquad (6)$$

$$\Delta G = -\ nFE \text{ at nonstandard conditions (6a)}$$

The Gibbs Free energy change is also related to the equilibrium constant, K, for a reaction, equation 7.

$$\Delta G^o = -\ 2.303RT \log K \qquad (7)$$

Equation 8 represents the relationship between K and E^o at 25^oC.

$$E^o = 0.0591/n \log K \qquad (8)$$

In this experiment , as a result of measuring the cell potential variation with temperature, it is also possible to ascertain several other relevant thermodynamic functions, ΔH (Heat of reaction), and ΔS(entropy-disorder). Mathematically , $\Delta G = \Delta H - T\,\Delta S$, equation (9), which is an equation of a straight line. If you plot ΔG versus T, the slope of the plot will equal $-\Delta S$, $\Delta S = -$ slope equation (10). If the equation , $\Delta G = \Delta H - T\,\Delta S$ is solved for ΔH, $\Delta H = \Delta G + T\Delta S$, ΔH can be calculated from the values of ΔG and ΔS. The units of ΔH and ΔG are kJ/mol, while that of ΔS is J/K . The sign of ΔH indicates whether a reaction liberates (gives off) heat , or whether it requires (absorbs) heat . A negative (-) value indicates heat liberation and is termed exothermic, while a positive (+) value indicates absorption and is termed an endothermic reaction.

Experimental Procedure (GROUP or CLASS EXPERIMENT)

A. Construction of electrochemical cells and measuring their potentials.

SALT BRIDGE PREPARATION

1. Prepare 250 mL of 0.1M KNO_3 (27.5g/250mL)
2. Boil 100mL of the salt solution, and remove it from the heat.
3. Add 1g of agar , stirring constantly.

4. Prepare 6 U-tubes and fill each one with warm KNO_3-agar
5. Dip and soak pieces of cotton into the 0.1M KNO_3 solution.
6. Remove with a tweezer and insert into the ends of the U-tubes, removing air bubbles. (This must be done while the KNO_3-agar is warm).

VOLTAIC CELL PREPARATION

You will construct the following electrochemical cells (voltaic cells) using Figure 15.1 as a guide. Hard glass test tubes may be substituted:

Voltaic Cells	Solution	Electrode	Solution	Electrode
Zn-Cu Cell	0.1M $Zn(NO_3)_2$	Zn	0.1M $Cu(NO_3)_2$	Cu
Pb-Cu Cell	0.1M $Pb(NO_3)_2$	Pb	0.1M $Cu(NO_3)_2$	Cu
Zn-Pb Cell	0.1M $Zn(NO_3)_2$	Zn	0.1M $Pb(NO_3)_2$	Pb
Sn-Pb Cell	0.1M $SnCl_2$	Sn	0.1M Pb $(NO_3)_2$	Pb

How do you make a Zinc-Copper Cell?

1. Clean the Cu and Zn electrodes with steel wool, place the Cu strip in a 150-ml beaker or hard glass test tube containing approximately 50 mL of 0.1M $Cu(NO_3)_2$, and the Zn strip into 50mL of 0.1M $Zn(NO_3)_2$ in another beaker or hard glass test tube.

2. Place a salt bridge into the half-cells.

3. Connect one clip to the Zn electrode and to the negative terminal,(-), (oxidation-half) of the voltmeter and connect another clip to the Cu electrode and to the positive terminal (+), (reduction-half). You now have a complete cell for measurement of the cell potential.

4. Read and record the voltage in volts for the cell and the temperature,^oC. If the potential is negative, reverse the electrodes.

5. Calculate ΔG(using $\Delta G = -nFE_{cell\ (measured)}$, and Calculate E^o_{cell} using the values in the table below, and then calculate the equilibrium constant, K from

$$E^o cell = \frac{0.0592}{n} \log K \ @25^oC$$

6. Repeat the above for the other electrode systems.

Reduction half reactions	E^o
Cu^{+2} + 2e- ---> Cu	+0.34v
Sn^{+2} + 2e- ---> Sn	-0.14v
Pb^{+2} + 2e- ---> Pb	-0.13v
Zn^{+2} + 2e- ---> Zn	-0.76v

B. EFFECT OF TEMPERATURE ON CELL EQUILIBRIA

1. Using the Sn-Cu cell , place it into an ice water bath in a 600mLbeaker. The bath will be heated as necessary for additional temperature measurements beyond room temperature.

2. When the temperature reaches 10°C , read and record the voltage from the voltmeter. Repeat at 20°C, 30°C, 40°C, 50°C.

C. EFFECT OF CONCENTRATION

In this experiment we will also see how cell voltages are related to the concentration of the components of the cell. The E cell for the redox reaction

$$Zn + Cu^{+2} \text{ ---------> } Cu + Zn^{+2}$$

will be measured at several known Cu^{+2} concentrations while maintaining the $[Zn^{+2}]$ constant . The Nernst equation is:

$E_{cell} = E^{o}_{cell} - 0.0591/2\ (\log [Zn^{+2}] / [Cu^{+2}])$ according to equation (5)

$$E_{cell} = E^{o}_{cell} - 0.0591/2 \log [Zn^{+2}] + 0.0591/2 \log [Cu^{+2}] =$$
$$y = b + m\,x$$

A plot of E_{cell} vs log [Cu^{+2}] yields an approximate slope, m of 0.0591/2 and a y intercept, b, of $E^{o}_{cell} - 0.0591/2 \log [Zn^{+2}]$.

1. Using 1M $Cu(NO_3)_2$ solution 1 , prepare solutions 2-4 using the following dilutions:

	Dilute to 100mL	Dilute to 100mL	Dilute to 100mL
Solution 1	-------------->Solution 2	-----------> Solution 3	-----------> Solution 4
1M $Cu(NO_3)_2$	1.0mL	1.0mL	1.0mL

   Each solution should be prepared in a 100mL volumetric flask using distilled water for dilution to the calibration marks on the flasks.

2. Calculate the Molarity of each solution.
3. Clean the Cu and Zn electrodes with steel wool, place a Zn strip in a 150-ml beaker or hard glass test tube containing approximately 50 ml 1M $Zn(NO_3)_2$, and the Cu strip into 50mL of Solution 2 in the other beaker or hard glass test tube.
4. Place a salt bridge into the half-cells.

5. Connect one clip to the Zn electrode and to the negative terminal, (-), (oxidation-half) of the voltmeter and connect another clip to the Cu electrode and to the positive terminal (+), (reduction-half).

6. Read and record the voltage in volts for the cell . If the potential is negative, reverse the electrodes.

7. Repeat for Solutions 3 and 4. Please clean and rinse the copper half-cell after each measurement.

8. Plot E_{cell} (y-axis) versus log $[Cu^{+2}]$ (x-axis) on a linear graph paper for four concentrations of $Cu(NO_3)_2$. and determine the slope.

9. Obtain an unknown Cu^{+2} solution, place 50mL of it in the copper half-cell compartment, and measure the cell potential .

10. From the standard curve determine the $[Cu^{+2}]$.

Sample calculations:

ΔG from **$E_{cell(measured)}$**. If in the Pb-Cu cell, the cell potential was measured and found to equal **+0.50v= $E_{cell(measured)}$**.

$Cu^{+2}(aq) + 2e^- \longrightarrow Cu(s)$, $E^o = +0.34v$
$Pb^{+2}(aq) + 2e^- \longrightarrow Pb(s)$, $E^o = -0.13v$

$Cu^{+2}(aq) + 2e^- \longrightarrow Cu(s)$, $E^o = +0.34v$
$Pb(s) \longrightarrow Pb^{+2}(aq) + 2e^-$ $E^o = +0.13v$

$Pb(s) + Cu^{+2}(aq) \longrightarrow Pb^{+2}(aq) + Cu(s)$, **Cell reaction, E^o cell = +0.47v**

Cell Diagram : Pb(s) | Pb^{+2}(0.1M) || Cu^{+2}(0.1M) | Cu(s)

ΔG = - nFEcell , F= 96,500 J/V-mol e^-

ΔG = - 2mol e^- x 96,500 J/V-mol e^- x **+0.50v** = -96,500J = **-96.5kJ**

$$E^o_{cell} = \frac{0.0592}{n} \log K \ @25^oC$$

E^ocell = +0.47v = 0.0592/2 x log K, +0.47 x 2/ 0.0592 = log K

log K = 15.8784, K = $10^{15.8784}$ = 7.56 x 10^{15}

Temperature:

If you plot ΔG versus T, the slope of the plot will equal **-ΔS, ΔS = - slope**

If the plot has a slope of -55.6J/°K, ΔS = - (-55.6J) =+ 55.6J/°K = **+0.0556kJ/°K**

ΔH= ΔG +TΔS = -96.5kJ + (298°K)x(+0.0556kJ) = -96.5kJ +16.57kJ= **-79.9kJ**

Concentration:

The voltage of a cell composed of 1M Zn^{+2} and an unknown Cu^{+2} solution was measured and found to be +0.99V at a log of $[Cu^{+2}]$ = -3.1 .

log $[Cu^{+2}]$ = -3.1, **$[Cu^{+2}]$** = antilog (0.9-4) = **7.9E-4M**

EXPERIMENT 15 ELECTROCHEMISTRY(GALVANIC CELLS)

DATA AND REPORT SHEET

NAME________________________________ASSIGNED #_____ DATE___ SCORE______
UNKNOWN#________

A. ELECTROCHEMICAL CELLS AND THEIR POTENTIALS.

Cell	Diagram	Anode Reaction	Cathode reaction	E^ocell calculated	Ecell Measured	Cell reaction
Zn-Cu						
	°C__________	Equil. Const. K=________	ΔG= _______			Spontaniety?_______

Cell	Diagram	Anode Reaction	Cathode reaction	E^ocell calculated	Ecell Measured	Cell reaction
Pb-Cu						
	°C__________	Equil. Const. K=________	ΔG= _______			Spontaniety?_______

Cell	Diagram	Anode Reaction	Cathode reaction	E^ocell calculated	Ecell Measured	Cell reaction
Zn-Pb						
	°C__________	Equil. Const. K=________	ΔG= _______			Spontaniety?_______

Cell	Diagram	Anode Reaction	Cathode reaction	E^ocell calculated	Ecell Measured	Cell reaction
Sn-Cu						
	°C__________	Equil. Const. K=________	ΔG= _______			Spontaniety?_______

Show ΔG and K Calculations for any one cell. Part A

B. EFFECT OF TEMPERATURE ON CELL EQUILIBRIA

Sn-Cu cell

Temperature °C	Temperature K	E_{cell}	$\Delta G=-nFE_{cell}$

Show one ΔG calculation at one temperature.

The Slope of the ΔG versus Temperature K= ____________ (Show Determination)

The value of ΔS = - slope = ________________________J/K.mol___ (Show Determination)

The value of ΔH = ΔG+TΔS = ________________________kJ___ (Show Calculation)

C. EFFECT OF CONCENTRATION

Unknown #_________

	Solution 1	Solution 2	Solution 3	Solution 4
Molarity, Cu^{+2}				
log $[Cu^{+2}]$				
E_{cell}				

Plot E_{cell} vs log $[Cu^{+2}]$

E_{cell} of Unknown Solution__________ $[Cu^{+2}]$_________ (Show Determination)

Questions:

1. Compare the sum of the Zn-Pb and Pb-Cu cell potentials to the Zn-Cu cell potential.

2. As the concentration of Cu^{+2} decreased, what is the observed trend if any between concentration and E_{cell}? Part C

3. Did the entropy, ΔS decrease, or increase in the Sn-Cu cell reaction as the reaction proceeded? Explain your answer.

4. Was the Sn-Cu cell reaction exothermic or endothermic?

5. What was the observed relationship if any between temperature and E_{cell} for the Sn-Cu cell?

6. Use your textbook to determine the cell voltage E^o cell , of a galvanic cell consisting of Sn^{+2}(1M)/Sn and Mg^{+2}(1M)/Mg.

Pre-Laboratory Assignment Experiment 15
Electrochemistry (Galvanic Cells) Name________________________________#____Score_____

Redox couples	E^o
$Br_2 + 2e^-$ ----------> $2Br^-$	+1.07v
$Fe^{+3} + 1e^-$---------> Fe^{+2}	+0.77v
$Sn^{2+} + 2e^-$ ---------> Sn	-0.14v
$Cr^{3+} + 3e^-$ --------> Cr	-1.0v

1. Consider the cell consisting of the Fe^{3+}(1M)/ $Fe^{2+}$and and Br_2(1atm)/ Br^-(1M) redox couples.
   a. Write the reduction half-reaction and the corresponding standard reduction potential and indicate which electrode does this occur at.

   b. Write the oxidation half reaction and the corresponding oxidation potential indicate which electrode does this occur at.

   c. Write the equation of the overall cell reaction and the value of E^o_{cell}?

2. Determine the cell voltage, E^o $_{cell}$, of a galvanic cell consisting of Sn^{+2}(1M) / Sn and Cr^{+3} (1M)/ Cr redox couples.

3. Use the Nernst Equation to determine the cell potential, E_{cell}, of the galvanic cell consisting of Sn^{+2}(0.0001M) /Sn and Cr^{+3} (0.1M)/Cr redox couples.

4. State the purpose of the salt bridge in the experiment ?

5. What is the sign of the anode in a galvanic cell? ______What electrochemical process occurs at the cathode?

10 DIVISIONS PER CENTIMETER

EXPERIMENT 16

THE RATE OF THE REACTION OF T-BUTYL CHLORIDE AND WATER

OBJECTIVES

a) To measure the rate of a chemical reaction and determine the rate law for the reaction
b) To examine the affect of reactant concentration and temperature on the rate of a reaction
c) To use a graphical method to calculate the activation energy and the specific rate constant for the reaction

SUPPLIES AND EQUIPMENT

2-chloro-2-methylpropane (t-butyl chloride, t-BuCl), Absolute Ethanol (dried over $MgSO_4$) Saturated NaCl solution, and pH buffers (pH = 7 or pH =4), Semilog Paper, Graph Paper, pH meter with a combination glass electrode, Water bath and thermometer, 150 mL Beaker, 10 and 100 mL Graduated cylinders, three Medicine droppers , a Magnetic stirrer (optional), a Timer or stopwatch reading time to the nearest second, Ice

WHAT YOU SHOULD KNOW BEFORE YOU BEGIN

Why can't you live to be 200 years old? It could be related to how long it takes for your muscles to lose their elasticity, i.e. it could be related to what we commonly call the ageing process. Why does an auto body begin to rust and look ugly after about 10 years? Its due to the reaction of iron with water and oxygen of the air. We know quite a lot about this latter reaction including how long it takes. These are two examples of relatively slow chemical reactions but fast reactions are of interest also to scientist. What we would like to do is to be able to measure the speed or rate of a chemical reaction and then to **control** that rate. To measure the rate of a chemical reaction we need only measure the amount of either a reactant or a product at various times after we allow the reactants to mix and the reaction to begin. For example the rate of disappearance of a reactant, M, would be:

$$\text{rate of disappearance M} = \frac{\text{change in concentration of M}}{\text{time required for the change}}$$

Fortunately, scientists also know the factors which affect the rate of a reaction, and by changing these factors we can change the reaction rate. For a given set of reactants, the concentration of the reactants, the temperature and the presence of a catalyst can all affect the reaction rate. For reactions occurring in solution, the nature of the solvent is also important. In today's experiment you will observe the effect of changing the concentration and the temperature on the rate of the reaction of 2-chloro-2-methylpropane, C_4H_9Cl, (t-BuCl) and water.

$$C_4H_9Cl \ + \ H_2O \ \text{----->} \ C_4H_9OH \ + \ HCl$$

Concentration effects- The effect of reactant concentrations are best expressed mathematically in equations we call rate laws. For a general reaction

$M + N \text{ ---->} C$ a possible rate law would be

rate = k $[M]^x$ $[N]^y$ where k is a constant (if temperature is constant), the specific rate constant.

X and Y are also constants, either fractions or whole numbers. The rate of a reaction, and the values of k, x, and y, must all be determined experimentally. The order of a reaction is the sum of x +y. If x=1 and y= 1 then the order of the reaction is two, or second order. We have equations relating the concentrations of reactants and time for the most common kinds of reactions, first and second order reactions. For a first order reaction, M -----> products,

$\ln [M_t] = \ln [M_o] - kt$ where $[M_o]$ is the initial molar concentration of M and $[M_t]$ is the molar concentration of M at some time t, after the reaction has begun. In common logarithms:

$\log [M_t] = \log [M_o] - kt/2.303$. A graphical plot of ln [M] versus t will yield a straight line with a slope equal to -k, the specific rate constant, or a plot of log [M] vs. t will yield a slope of -k/2.303.

For a second order reaction, M ----> products

$1/[M_t] - 1/[M_o] = kt$ Here a plot of $1/[M_t]$ versus time yields a straight line with a slope of k.

Temperature effects- Generally reactions are faster at higher temperatures. The rate of a reaction approximately doubles for each 10 K increase in temperature. Again the effect of temperature on reaction rate is best expressed mathematically in the Arrhenius equation.

$$\text{rate (or specific rate constant)} = A\, e^{-Ea/RT}$$

where A is a constant (which can have values of almost 0 to 1), R is the gas constant = 8.314 J/mol K, T is the absolute temperature and Ea is the activation energy. The activation energy is the minimum energy required to break "old bonds" so that new bonds can form. Reactions with low activation energies will proceed much faster than reactions with high activation energies. Rearranging the Arrhenius equation yields;

$$\log (\text{rate or } k) = - Ea/2.303\, RT + \log A$$

and a plot of log k (or log rate) versus 1/T will yield a straight line with a slope of -Ea/2.303R.

For today' s experiment we will study the reaction

$$C_4H_9Cl + H_2O \text{ -----> } C_4H_9OH + HCl$$

and measure the rate of the reaction by measuring the concentration of HCl, one of the products. Of course HCl is an acid. How would you measure the concentration of a strong acid like HCl in a solution? You could measure the pH. Every mole of C_4H_9Cl produces a mole of HCl and each mole of HCl produces a mole of H^+ in solution. So as we measure the decrease in solution pH we'll be measuring the decrease in concentration of C_4H_9Cl. How do we calculate the $[C_4H_9Cl]$ at various times? At the completion of the reaction the $[H^+]$ will be equal to the initial concentration of C_4H_9Cl, $[C_4H_9Cl]_{initial}$. The pH at a given time will yield the $[H^+]$ at that time which is also the $[C_4H_9Cl]$ consumed at that time. Therefore the $[C_4H_9Cl]$ remaining is $[C_4H_9Cl]_{initial}$ - $[C_4H_9Cl]$. A table summarizing these relationships is

Time (sec)	pH	$[H^+]$	$[C_4H_9Cl]_{initial}$	$[C_4H_9Cl]$ remaining
19.9	4.02	9.55 E-5	3.24 E -4	(3.24 E -4) - 9.55 E -5
29.6	3.63	2.34 E-4	"	(3.24 E -4) - 2.34 E -4
49.4	3.49	3.24 E -4	"	(3.24 E-4) - 3.24 E -4 = 0
54.1	3.49	3.24 E -4	"	0

As you can see this is a fast reaction and this allows you to start the reaction numerous times, each time changing either the concentration of the reactant or the temperature.

EXPERIMENTAL PROCEDURE

1. Calibrate your pH meter using two standard pH buffers (pH 4 and 7).

2. Measure the temperature of your buffers and use this value to properly set the temperature compensation dial of the pH meter.

3. Using a 100 mL graduated cylinder, measure 100 mL of deionized water and transfer to a clean 150 mL beaker; add three drops of the saturated NaCl solution to stabilize the initial pH reading of the mixture.

4. While stirring the solution record the initial pH reading of the solution and the temperature.

5. Now, while stirring, quickly add six drops of pure t-butyl chloride and begin timing after the addition of the last drop. Record the pH and the time (in seconds) for about three minutes or until the pH readings become constant. Be sure to continually stir during the reaction and to make sure that the pH electrode is well immersed in the solution. Use your judgement about how often to take a reading, but remember the reaction will be fast early and will slow down later on. You should take at least a dozen readings.

6. A freshly prepared solution is needed for each of the following determinations.

Concentration effects- Table 1

7. Prepare the water-saturated NaCl solution as above.

8. To prepare a 75% solution of t-butyl chloride using your 10 mL graduated cylinder, mix 3 mL of t-butyl chloride and 1 mL of ethanol. When ready to begin the reaction, add six drops of this mixture to the water-saturated NaCl solution and begin recording the pH and time as above.

9. Prepare a 50% solution of t-butyl chloride by mixing 3 mL of t-butyl chloride and 3 mL of ethanol. Repeat the procedure above.

Temperature effects- Table 2

11. Prepare the 50% solution of t-butyl chloride as described above.

12. This time add three drops of the mixture to the water-saturated NaCl mixture and take pH readings for 3 minutes at 25 °C. Repeat the experiment at 10, 15 and 20 °C. Be sure to keep the temperature constant while taking the pH readings.

References

Brown, William H. "The Rate of a Chemical Reaction: The Solvolysis of tert-Butyl Chloride", in Modular Laboratory Program in Chemistry; Grant: Boston, 1972.

Anthony Allen, et. al., J. Chem. Ed. 68 (7), 609 (1991).

Roland P. Stout, Chem 13 News, p 19, April/May 1987.

EXPERIMENT 16 THE RATE OF THE REACTION OF T-BUTYL CHLORIDE AND WATER

DATA AND LAB REPORT SHEET

NAME______________________ASSIGNED #______DATE________SCORE______

CALCULATIONS

1. Use your 100% t-butyl chloride to determine the reaction order.

2. Complete the entries in Table 1 of the Data Sheet. Then make two graphs.

3. On one plot log [C_4H_9Cl] versus t, on the other plot 1/[C_4H_9Cl] versus t.

4. Which plot yields a straight line?

5. From Table 1 and Graph 1

   Order of the reaction________________

   Slope at 25°C______________________________(Show Calculation)

6. From the slope of the graph calculate the specific rate constant, k. (Show Calculation)

7. Plot the 100% , 75% and 50% t-butyl chloride results (concentration of t-butyl chloride versus time) .

8. How does changing concentration affect the rate of this reaction?

9. Complete the entries in Table 2. Using the graphical technique described above calculate the specific rate constants at the four different temperatures. Show calculations for at least one.

10. Use these specific rate constants to complete Table 3.

Table 3. Values of k, Specific Rate Constant at Various Temps.

Temp (K)	1/Temp (1/K)	k	log k

11. Finally on separate graph paper plot log k versus 1/T .

12. From Table 3 and Graph 3
Slope of log k versus 1/T ________________________(Show Calculation)

13. Calculate the Activation Energy, Ea. (Show Calculation)

RATE OF REACTION OF t-Butyl Chloride DATA SHEET No. 1

Name__________________________ Expt. 16

Table 1. Effect of Concentration Temperature____________________

Initial % of t-BuCl	Time(s)	pH	$[H^+]$	[t-BuCl]	log[t-BuCl]	1/[t-BuCl]
100						
75						
50						

RATE OF REACTION OF t-Butyl Chloride DATA SHEET No.2

Name___________________________ Expt. 16

Table 2. Effect of Temperature 50% t-butyl chloride

T(K)	Time(s)	pH	$[H^+]$	[t-BuCl]	log[t-BuCl]	1/[t-BuCl]
283K						
288K						
293K						

Pre-Laboratory Assignment Experiment 16
The Rate of The Reaction of t-Butyl Chloride and Water
Name________________________________#________ Score________

1. In today's experiment a student measures a pH of 3.55 after 40 seconds. The pH stabilizes and becomes constant at 3.10 after 70 sec.
   a) What is the final [H^+] ?

   b) What is the initial conc. of t-butyl chloride?

   c) What is the conc. of t-butyl chloride remaining after 40 seconds?

2. What is the purpose of calibrating the pH meter with two different buffers instead of just one buffer?

3. If we use the Arrhenius equation to express the specific rate constant in terms of E_a and T, the rate of reaction of C_4H_9Cl is: rate $= A\, e^{-Ea/RT}$ [C_4H_9Cl]
Circle the following variables which have an effect on the rate:

a) [C_4H_9Cl] b) the size of the beaker in which the reaction occurs c) temperature
d) the volume of deionized water added to the 150 mL beaker
e) continuous stirring

4. Describe **all** the experimental conditions you would need to slow this reaction down as slow as possible in our labs.

5. Given: Complete the table and plot log of rate vs 1/T(1K). Attach the plot. Determine the slope and calculate the Energy of Activation, E_a, E_a= -2.303Rx slope. R= 8.314J/K.mol

Temp C	T(K)	1/T(1/K)	Rate(L/mol.s)	log rate
5			1.92E-4	
15			2.60E-4	
25			3.42E-4	
35			4.43E-4	

10 DIVISIONS PER CENTIMETER

10 DIVISIONS PER CENTIMETER

10 DIVISIONS PER CENTIMETER

10 DIVISIONS PER CENTIMETER

EXPERIMENT 17

Le CHATELIER'S PRINCIPLE

OBJECTIVES

a) To understand and apply Le Chatelier's Principle to equilibria
b) To write ionic equations involving equilibrium shifts

SUPPLIES AND EQUIPMENT

0.1M $BaCl_2$, 1M K_2SO_4, 6M H_2SO_4, 6M NaOH, 1M K_2CrO_4, 6M HCl, 3M HCl, 1M NH_3(aq), 0.1M NH_3(aq), 1M NaOAc, 1M NH_4Cl, Phenolphthalein(aq), Methyl Orange(aq), Methyl violet(aq), 0.1M $NiCl_2$, Solid $CoCl_2{\cdot}6H_2O$, Test tubes, Ring stand

WHAT YOU SHOULD KNOW BEFORE YOU BEGIN

Le Chatelier's Principle was developed by Henri Louis Le Chatelier in the late 1800's. According to the concept, " **If a reaction that is in a state of dynamic equilibrium is disturbed (stressed), the reaction will shift in a direction to counteract the disturbance and thus minimizing the stress".** On a daily basis, you encounter incidences of the application of Le Chatelier's Principle. e.g. You eat because your brain signals to the body that it is time to provide it with certain nutrients, and the body says it's time to eat, and therefore to fill the void, you eat. The principle is applicable to numerous events in nature. There are many such events that this principle is applicable to. Can you think of at least three ? The constants that govern chemical reaction equilibria are identified by a number of K values, e.g. K_e, K_b, K_c, K_{sp}, etc. These constants have actual numerical values. The numerical values are constant if the concentrations of the involved species are changing. However, the constants do have a temperature dependence. Many reactions essentially do go to completion, however , there are some which stop and form equilibrium mixtures between the reactants and products. They reach a point in which the rate of formation of products equals the rate at which reactants form , which is a state of dynamic equilibrium.

Once equilibrium has been established, the composition of equilibrium mixtures can be influenced, or changed by the addition or removal of reactants or products or by other factors such as temperature change. What effect does temperature have on the equilibria? To understand the effect of temperature on equilibria, first we should note whether the reaction in question is endothermic or exothermic, as many reactions evolve or require heat addition to occur. Using the following equilibrium,

$2SO_2(g) + O_2(g) \Longleftrightarrow 2\,SO_3(g)$, $\Delta H = -197kJ$, the reaction is an exothermic reaction. When heat is added to an exothermic reaction, the reaction shifts to absorb the added heat , which means that shift to the left (reverse) would occur. If heat is removed, the reaction shifts to replace the heat that was removed , which means a shift to the right (product). The reverse occurs for endothermic reactions. The equilibrium constant expression for the equilibria is:

$$Ke = \frac{[SO_3]^2}{[SO_2]^2[O_2]} .$$

If the concentration of a reactant or product changes, the reaction will shift to counteract the disturbance. If the partial pressure of SO_3 is increased, the reaction will go to the left thus increasing the reactants , SO_2 and O_2.

$$2SO_2(g) + O_2(g) <===> 2\ SO_3(g)$$
$$< \text{------ increase } SO_3$$

If the partial pressure of O_2 is increased, the reaction will shift to the right thus increasing the SO_3 and causing the SO_2 to decrease to re-establish equilibrium.

$$2SO_2(g) + O_2(g) <===> 2\ SO_3(g)$$
$$\text{increase } O_2\text{----->} .$$

If chloride ion is added to an aqueous solution of Ag+ , a white solid of AgCl will form. The reaction equilibria is represented by the following equation:

$Ag^+(aq) + Cl^-(aq) <===> AgCl(s)$

The addition of ammonia to the above equilibria will cause the reaction to shift to the left (reverse) according to Le Chatelier's Principle to replace the silver ion which is removed.
$Ag^+(aq) + Cl^-(aq) <===> AgCl(s)$

The ammonia will remove silver ions forming the coordination complex called Diamminesilver(I) ion: $[Ag(NH_3)_2^{+1}]$.
$Ag^+(aq) + 2NH_3(aq) <===> Ag(NH_3)_2^{+1}$

The shift in equilibrium as a result of the addition of an ion common to the equilibrium mixture is known as the Common Ion Effect. Adding Cl^- to the equilibrium above decreases the $[Ag^+]$, and increases the solid AgCl. The equilibrium has shifted to the right. In this experiment, the common ion effect will be utilized as we study the reaction equilibria.

EXPERIMENTAL PROCEDURE

A. CONCENTRATION EFFECTS

The reaction of barium chloride with the potassium sulfate yield the formation of a white solid of barium sulfate.

1. Add 10 drops of 0.1M barium chloride to 5mL of 1M potassium sulfate, in a test tube. Note and record all observations.

2. Write the ionic equation of the reaction equilibria.

3. Add 15drops of 6M HCl to the test tube , note and record all observations. What has occurred? Explain what has happened using ionic equations, and Le Chatelier's Principle.

The reaction of the yellow chromate [CrO_4^{-2}] ion with the hydrogen ion from an acid to yield the orange dichromate ion $Cr_2O_7^{-2}$.

$2CrO_4^{-2} + 2H^+ \Longleftrightarrow Cr_2O_7^{-2} + H_2O$

4. Add 10 drops of 6M sulfuric acid to 2mL of 1M potassium chromate in a test tube. Note and record all observations.

5. Add 5drops of 6M NaOH to the test tube, note and record all observations. What has occurred? Explain what has happened using ionic equations.

6. Re-acidify the mixture by adding 6M sulfuric acid ,note the result and explain.

The reaction of nickel chloride with ammonia yields the formation of a coordination complex named: hexaamminenickel (II) ion.

$Ni(H_2O)_6^{+2}(aq) + 6NH_3(aq) \Longleftrightarrow Ni(NH_3)_6^{+2} + 6H_2O$

green

7. Add 15 drops or more of 1M NH_3 to 15drops of nickel chloride until a change in color is noted.

8. Add drops of 3M HCl until a color change occurs. Note and record all observations. Explain using the reaction using the ionic equation.

The reaction of the transition metal ion hexaaquacobalt(II) ion (octahedral) with chloride ion resulting in a change in the geometry to the tetrachlorocobalt(II) ion (tetrahedral). Transition metals often have different geometries depending upon the extent of coordination (bonds to atoms or ions). Different numbers of bonds will yield different geometries.

$Co(H_2O)_6^{+2}(aq) + 4Cl^{-1}(aq) \Longleftrightarrow CoCl_4^{-2} + 6H_2O$

9. Add 3 small crystals of hexaaquacobalt(II) chloride to a test tube . Add 2mL of 6M HCl and stir to dissolve the solid. Note and record all observations.

10. Add water slowly dropwise , stopping to stir the solution, until the color of the solution no longer changes.

11. Place the test tube in a beaker of boiling water. Note and record all observations.

12. Cool the test tube under running water, and note all observations. Explain what has occurred.

Acid Base indicators are themselves weak acids or weak bases which are usually organic compounds that undergo color changes in acid-base equilibria. The color is different dependent upon whether the medium is acidic or basic. The acid form is designated as HIn, while the conjugate base form is designated as In^- in general. The reaction of the equilibria is given as:

$HIn \Longleftrightarrow H^+ + In^-$

Methyl red indicator is red in the HIn form and yellow in the In^- form.

$HIn \Longleftrightarrow H^+ + In^-$

Red Yellow

The formulas of each form is shown below:

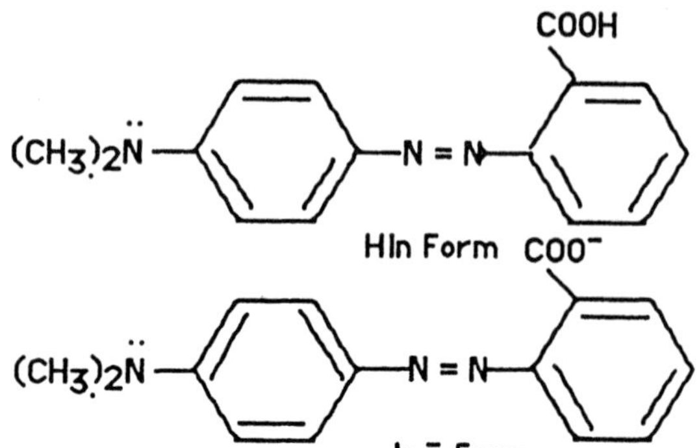

Methyl Red

In^- Form

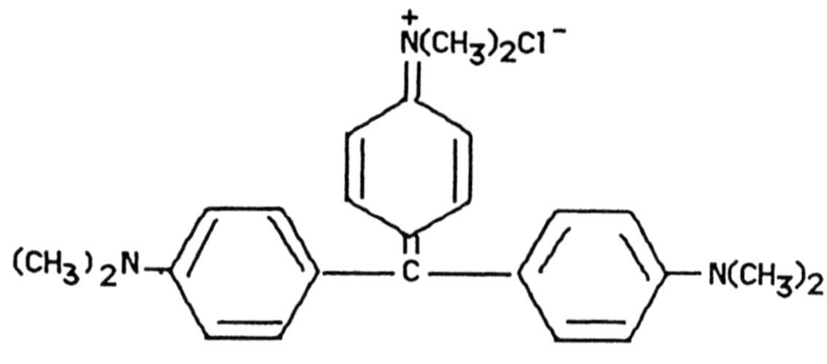

Methyl Violet

The indicators change colors at different H^+ ion concentrations. Phenolphthalein is another such indicator which is colorless in the HIn form and pink(reddish) in the In- form.

Strong Acid-Base Effect (Acid-Base Indicators)

13. Add 2 drops of methyl violet to 2 ml of water in a test tube, and then add 2 drops of 6M NaOH, note and record all observations. What has occurred? Write an equation to explain what has occurred?

14. Add to the same test tube drops of 6M HCl, note and record all observations. What has occurred? Write an equation to explain what has occurred?

Weak Acid-Base Effect

Acetic acid equilibria $HOAc \Longleftrightarrow H^+ + OAc^-$

15. Add 2 mL of 0.1M HOAc to 2 test tubes. Add 1 drop of methyl orange to each. Note and record observations.

16. Add dropwise 1M NaOAc (Common Ion) to the second test tube until a change occurs. The first test tube is prepared for color comparison. What has occurred?

17. Add 2 ml of 0.1M NH_4OH to two test tubes. Add 1 drop of Phenolphthalein to each. What has occurred? Note and record the color.

18. Add 1M NH_4Cl dropwise to one test tube, stirring after each drop. What do you observe? Can you detect any unusual odor? Write the ionic equation. Explain your observations.

19. Add dropwise 6M HCl to the other test tube from #17 above. What is occurring? Write the equation for the reaction of NH_3 with water where $NH_4^+$and OH^- are formed. Explain the results in terms of the NH_3 , NH_4^+ equilibria.

Precaution: 1. Properly dispose of the chromium salts in the waste container.

EXPERIMENT 17 Le CHATELIER'S PRINCIPLE
DATA AND LAB REPORT SHEET

NAME______________________________ASSIGNED #_____ DATE______ SCORE______

CONCENTRATION EFFECTS

1. Barium chloride + Potassium sulfate
   Observations____

2. Net Ionic Equation____
3. H^+ Addition
   Observations____

   Net Ionic Equation____

   Explain in terms of Le Chatelier's Principle______________________________

   ____
4. Sulfuric acid+ Potassium chromate(Color_____________?)

   Observations____

   Net Ionic Equation____

5. OH^{-1} Addition
   Observations____

   Net Ionic Equation____

   Explain in terms of Le Chatelier's Principle______________________________

   ____

6. H^{+1} Addition
   Observations____

   Net Ionic Equation ____

   ____
7. $NiCl_2 + NH_3(aq)$
   Observations____

   Net Ionic Equation____
8. H^{+1} Addition
   Observations____

   d)Net Ionic Equation____

   Explain in terms of Le Chatelier's Principle______________________________

9. $Co(H_2O)_6^{+2} + Cl^{-1}$
Observations_______________

Color of $Co(H_2O)_6^{+2}$_______________

Color of $CoCl_4^{-2}$_______________

Net Ionic Equation_______________

11. Heating the solution
Observations_______________

Explain in terms of Le Chatelier's Principle_______________

____

12. Cooling the solution
Observations_______________

Explain in terms of Le Chatelier's Principle_______________

____

The reaction is exothermic or endothermic? ___ Why?_______________

13. Methyl Violet + Water
Observations_______________

Net Ionic Equation (Use HIn /In-)_______________

14. OH^{-1} Addition
Observations_______________

Ionic Equation_______________

15. HOAc + Methyl Orange
Observations_______________

Net Ionic Equation_______________

16. OAc^{-1} Addition (Common Ion Effect)
Observations______________________________

Net Ionic Equation______________________________

17. $NH_3(aq)$ + Phenolphthalein
Observations______________________________

Net Ionic Equation______________________________

18. NH_4^{+1} Addition
Observations______________________________

Net Ionic Equation______________________________

19. H^{+1} Addition
Observations______________________________

Net Ionic Equation______________________________

QUESTIONS

1. What are the colors of each of the following?

$Cr_2O_7^{-2}$__________ $BaSO_4$__________

2. What is the Common Ion Effect?

3. If a reaction is exothermic, and we add heat to the reaction in which direction should it shift?

Pre-Laboratory Assignment Experiment 17
Le Chatelier's Principle Name_________________________________#______Score______

1. Write a net ionic equation for the precipitation of lead(II) chromate from aqueous solutions of its ions.

2. According to Le Chatelier's Principle if the concentration of lead ion is reduced, the reaction should :

3. Write the ionic equation for the formation of the deep blue $Cu(NH_3)_4^{+2}$ from the reaction of $Cu(H_2O)_4^{+2}$ (sky blue) with ammonia.

4. When Hydrogen ion is added to the solution, it turns sky blue. Write an equation for the reaction of H^+ with ammonia to explain the observation.

5. $2Ag^+ (aq) + CO_3^{-2} (aq) \Longleftrightarrow Ag_2CO_3(s)$

   H^+ addition : $2H^+ + CO_3^{-2} \longrightarrow H_2CO_3(aq)$

   Explain the effect of the addition in terms of Le Chatelier's Principle.

EXPERIMENT 18

EQUILIBRIUM CONSTANT-TRANSITION METAL COMPLEX-K_{eq}

OBJECTIVES

a) To determine the equilibrium constant for a reaction.
b) To pprepare a calibration curve and apply it.
c) To become familiar with the use of a spectrophotometer.
d) To understand the concept of absorbance and transmittance.

SUPPLIES AND EQUIPMENT

Spectronic 20, burets or pipets, 0.1M HNO_3, 2.00E-3M $Fe(NO_3)_3$ in 0.1M HNO_3, 2.00E-3M KSCN in 0.1M HNO_3, 0.2M $Fe(NO_3)_3$ in 0.1M HNO_3, Cuvets, Distilled Water, Kimwipes, Graph Paper, 100mL Volumetric Flasks

WHAT YOU SHOULD KNOW BEFORE YOU BEGIN

When light (white) is passed through a sample that is colored in solution, some of the light may be absorbed , while some of the light may be transmitted (pass through). The absorbed radiation is a result of the energy difference between two levels in the solute molecules. The molecule has become excited. The radiation that is not absorbed is thus transmitted and is detected by an instrument called a spectrophotometer. The relationship between the energy and the wavelength is : $E = hc/\lambda$ where h is Planck's Constant, c is the Speed of Light, and λ is the wavelength. A spectrophotometer is an instrument that is designed to measure the absorption of light at specific wavelengths. The absorption will occur at different wavelengths of light. The colored region of the electromagnetic spectrum which deals with this phenomenon is called the visible region. The values of the wavelength ranges from approximately 400nm (4000Å) to 700nm (7000Å) . The extent of absorption is proportional directly to the concentration of the absorbing species. The absorption maximum can be determined and utilized for numerous analyses. A diagram of a Spectrophotometer is shown below:

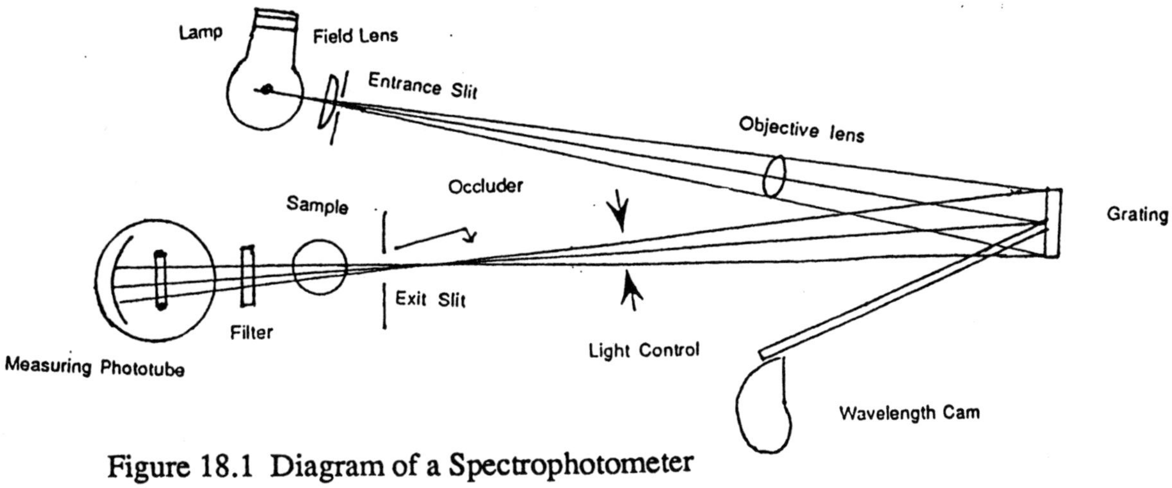

Figure 18.1 Diagram of a Spectrophotometer

In Table 18.1 is listed the colors of the absorbed light , colors of the transmitted light (complimentary- seen) , and the approximate wavelengths. The light that is absorbed is called the spectral color and the light that is seen is the light that is transmitted. When for example red light is removed from white light (absorbed) , the eye sees the color blue-green. A typical Spectrophotometer that is often used is the Spectronic 20 or Spectronic 20D. The spectronic 20 uses a tungsten -filament as the light source powered by a power supply which yields a constant radiation intensity. A simple grating is used for diffraction. The radiation passes through the sample to a phototube. The electrical signal generated from the detector is amplified which is then used to power a meter which has a calibration scale in absorbance or transmittance. A cuvet containing the sample

Table 18.1. Visible Region Spectral Color and Complimentary Color

Wavelength (nm)	Spectral (Absorbed)	Complimentary (Transmitted)
410	Violet	Yellow
430	Indigo	Yellow
480	Blue	Orange
500	Blue-Green	Red
530	Green	Purple
560	Yellow-Lemon	Violet
580	Yellow	Indigo
610	Orange	Blue
680	Red	Blue-Green

solution is placed in the instrument for analyses. The cuvet is usually of 1 cm cell length. The extent of absorption is determined by several factors. Included among these are (1) the concentration of the absorbing species, (2) the cell thickness -width of the sample cuvet, and (3) the molar extinction coefficient - probability of light absorption.

Absorbance (A) = εbc, ε = Molar extinction coefficient, b= cell path length, c = concentration (moles / liter) . This is an expression of Beers Law.

The Molar extinction coefficient is a constant at any wavelength. If the same spectrophotometer and the same wavelength are utilized, then the absorbance will be directly proportional to the concentration. Additionally , the Absorbance is mathematically equivalent to the negative logarithm of the Transmittance, T.

$A = -\log T = \log 1/T$. The instrument often gives the % Transmittance which would be need to be changed to T by moving the decimal 2 places to the left, e.g. 80%T = 0.8T.

In this experiment, a standard curve will be prepared using absorbance measurements of known concentration solutions of $FeSCN^{+2}$. The standard curve will then be used to determine the concentrations of $FeSCN^{+2}$ in a set of unknown equilibrium solutions. The equilibria that will be investigated is that of the reaction of Fe^{+3} with SCN^{-1} to form $FeSCN^{+2}$, a transition metal complex.

The equation is: $Fe^{+3} + SCN^{-1} \Longleftrightarrow FeSCN^{+2}$

$$\text{The equilibrium constant is : } K_{eq} = \frac{[FeSCN^{+2}]}{[Fe^{+3}][SCN^{-1}]}$$

The complex formed is a blood-red complex. Mixtures of the reactants will be prepared and the absorbance measured at a wavelength of 447nm. The reactions are carried out in 0.1M HNO_3 to prevent hydrolysis. A standard curve will be prepared of Absorbance vs [$FeSCN^{+2}$]. Unknown solutions will be prepared and the [$FeSCN^{+2}$] determined at reaction equilibrium from the standard curve. The [Fe^{+3}] initially is 2E-3M, and [SCN^{-1}] is 2E-3M. From the equilibrium concentration of $FeSCN^{+2}$ the amount of Fe^{+3} and the amount of SCN^{-1} at equilibrium will be determined, and the equilibrium constant calculated.

$$\text{The } [Fe^{+3}] \text{ at equilibrium} = \frac{\text{initial moles of } Fe^{+3} - \text{equilibrium moles of } FeSCN^{+2}}{\text{Volume Total in liters}}$$

$$\text{The } [SCN^{-1}] \text{ at equilibrium} = \frac{\text{initial moles of } SCN^{-1} - \text{equilibrium moles of } FeSCN^{+2}}{\text{Volume Total in liters}}$$

The equilibrium moles of $FeSCN^{+2}$ is determined by multiplying the Volume (liters) x [$FeSCN^{+2}$] at equilibrium.

The initial moles of reactants is determined by : V (liters) x Molarity (moles/liter)

EXPERIMENTAL PROCEDURE
(Designed for groups)
A. STANDARD CURVE

1. Obtain a clean buret, rinse it with distilled water and with 2.00E-3M KSCN (prepared in 0.1M HNO_3), and fill it with 2E-3M KSCN .

2. Obtain a second clean buret, rinse it with distilled water and with 2E-1M $Fe(NO_3)_3$ (prepared in 0.1M HNO_3), and fill it with 2E-1M $Fe(NO_3)_3$.

3. Place approximately 125mL of 0.1M HNO_3 in a 250mL beaker. Use for dilution to the calibration mark in each volumetric flask

4. Obtain six 100mL Volumetric flasks and label from #1-#6. Dispense from the buret the specified quantities as indicated in Table 18.2 below and dilute each to the 100mL calibration mark with 0.1M HNO_3.

CAUTION! Please label the burets carefully, use the correct Fe^{+3} for each part of the experiment.

Table 18.2. Standard Mixtures

Flask #	KSCN (0.002M)	$Fe(NO_3)_3$ 0.2M
1	0mL	25mL
2	1mL	25mL
3	3mL	25mL
4	5mL	25mL
5	7mL	25mL
6	9mL	25mL

*dilute to 100mL volume with $0.1MHNO_3$, the Fe^{+3} is 0.2M

5. Pour appropriately each sample into a cuvette, measure and record the absorbance/ transmittance of each sample using the Spectrophotometer. The use of which will be explained by the laboratory instructor.

6. Plot on linear graph paper the Absorbance vs [$FeSCN^{+2}$] and draw the best straight line through the data points.

B. EQUILIBRIUM UNKNOWN SOLUTIONS

1. Obtain a clean buret, rinse it with distilled water and with 2E-3M $Fe(NO_3)_3$ (prepared in 0.1M HNO_3), and fill it with 2E-3M $Fe(NO_3)_3$.

2. Obtain a clean buret, rinse it with distilled water and with 0.1M HNO_3, and fill it with 0.1M HNO_3.

3. Obtain five test tubes and label from #1-#5.
   Buret in the specified quantities as indicated in Table 18.3.

Table 18.3. Equilibrium Unknown Solutions

Test Tube #	KSCN	*$Fe(NO_3)_3$ 0.002M	HNO_3
1	5mL	4mL	1mL
2	4mL	4mL	2mL
3	3mL	4mL	3mL
4	2mL	4mL	4mL
5	1mL	4mL	5mL

* Fe^{+3} is 2.00E-3M

4. Pour appropriately each sample into a cuvet, measure and record the absorbance/ transmittance of each sample using the Spectrophotometer.

5. Using the standard plot, determine and record the value of [$FeSCN^{+2}$] at equilibrium for each unknown.

Precautions: 1. The blank solution contains the solvent system utilized in the experiment.
2. Always rinse the cuvet with the next sample to be analyzed.
3. Always wipe the cuvet to remove all foreign substances including finger prints before insertion into the Spectrophotometer

Sample Keq Calculations

e.g. The absorbance of unknown 1 solution was found to be 0.3
Using the standard curve the [$FeSCN^{+2}$] was found equal to 0.7E-4M

Moles of $FeSCN^{+2}$ at equilibrium = Volume (total) x [$FeSCN^{+2}$] =
0.01liters x 0.7E-4M = 0.7E-6moles

Moles of Fe^{+3} initially = V x M = 4E-3liters x 2E-3M = 8E-6moles

Moles of SCN^{-1} initially = V X M = 5E-3liters x 2E-3M= 10E-6moles

Moles of Fe^{+3} complexed = Equilibrium moles of $FeSCN^{+2}$
0.7E-6moles

Moles of SCN^{-1} complexed = Equilibrium moles of $FeSCN^{+2}$
0.7E-6moles

Moles of Fe^{+3} at equilibrium = Initial - Equilibrium moles
8E-6 moles - 0.7E-6moles = 7.3E-6moles

Moles of SCN^{-1} at equilibrium = Initial - Equilibrium moles
10E-6 moles - 0.7E-6moles = 9.3E-6moles

[Fe^{+3}] at equilibrium = Moles at equilibrium / V total = 7.3E-6moles/ 0.01liters= 7.3E-4M

[SCN^{-1}] at equilibrium = Moles at equilibrium / V total = 9.3E-6moles/ 0.01liters= 9.3E-4M

[$FeSCN^{+2}$] = 0.7E-4M

$$\text{Equilibrium Constant} = K_{eq} = \frac{[FeSCN^{+2}]}{[Fe^{+3}][SCN^{-1}]}$$

Keq= 0.7E-4M/ 7.3E-4M x 9.3E-4M = 103 M^{-1}

EXPERIMENT 18 EQUILIBRIUM CONSTANT-TRANSITION METAL COMPLEX

DATA AND LAB REPORT SHEET

NAME______________________________ASSIGNED #_____ DATE______ SCORE______

A. STANDARD CURVE

$[Fe^{+3}] = 0.2M$ $[SCN^{-1}] = 2.00E\text{-}3M$

Standard Mixtures

	1	2	3	4	5	6
Vol- SCN^{-1}	_______	_______	_______	_______	_______	_______
Moles-SCN^{-1}	_______	_______	_______	_______	_______	_______
$[SCN^{-1}]$	_______	_______	_______	_______	_______	_______
$[FeSCN^{+2}] = [SCN^{-1}]$	_______	_______	_______	_______	_______	_______
Absorbance, A	_______	_______	_______	_______	_______	_______

B. EQUILIBRIUM UNKNOWN SOLUTIONS

$[Fe^{+3}] = 2.00E\text{-}3M$ $[SCN^{-1}] = 2.00E\text{-}3M$

Unknown Mixtures

	1	2	3	4	5
Vol- Fe^{+3} liters	_________	_________	_________	_________	_________
Moles-Fe^{+3} initially	_________	_________	_________	_________	_________
Vol- SCN^{-1} liters	_________	_________	_________	_________	_________
Moles-SCN^{-1} initially	_________	_________	_________	_________	_________
Absorbance, A	_________	_________	_________	_________	_________

Determinations:

$[FeSCN^{+2}]$ Standard Curve					
Moles-$FeSCN^{+2}$ at equilibrium					
Moles of Fe^{+3} = Moles of $FeSCN^{+2}$ complexed					
Moles of Fe^{+3} uncomplexed					
$[Fe^{+3}]$ at equilibrium					
Moles of SCN^{-1} = Moles $FeSCN^{+2}$ complexed					
Moles of SCN^{-1} uncomplexed					
$[SCN^{-1}]$ at equilibrium					
K_{eq}= Equilibrium Constant					

K_{eq} Average
Equilibrium Constant ____________________

K_{eq} Standard Deviation ____________________

Show Determinations on lineless paper completely for any one set (column).

1. If the Absorbance of $FeSCN^{+2}$ was read higher than what it should be , what effect if any would this have on the value of Keq? explain

Pre-Laboratory Assignment Experiment 18
Equilibrium Constant Name______________________________#______Score______

1. Write the equation and the equilibrium constant expression for the reaction under study in the experiment.

2. What species is being spectrophotometrically measured?

3. What is the color of the species?

4. What is a standard curve?

5. Given the following data prepare a Plot of Absorbance vs [$FeSCN^{+2}$]

Absorbance	[$FeSCN^{+2}$] x10^{-4}
0.01	0.02
0.2	0.4
0.5	1.1
0.71	1.59

6. What is the [$FeSCN^{+2}$] at an absorbance of 0.35 from the standard curve ?

10 DIVISIONS PER CENTIMETER

10 DIVISIONS PER CENTIMETER

EXPERIMENT 19

IONIZATION CONSTANT OF A WEAK ACID, K_a

OBJECTIVES

a) To learn how to properly use a pH meter.
b) To become acquainted with buffers.
c) To determine an ionization constant, K_a for a weak acid.
d) To calculate the % ionization of the acid.

SUPPLIES AND EQUIPMENT

0.60 M unknown acid*, 0.30 M NaOH, pH-meter, Buffers-pH 4.0,and 7.0, Burets, 25mL Graduated pipets, Graph Paper , 100mL Volumetric Flasks
*(The unknown acids could be acetic acid, formic acid, lactic acid, etc.)

WHAT YOU SHOULD KNOW BEFORE YOU BEGIN

An acid, according to Bronsted-Lowry, is any substance that is a proton donor. The Bronsted-Lowry base is any substance that is a proton acceptor. Therefore, a weak acid, HB, in aqueous solution will have an acid-base reaction with water; water acting as a Bronsted-Lowry base:

$$HB + H_2O = H_3O^+(aq) + B^-(aq) \qquad (1)$$

This reaction is an equilibrium reaction, meaning that only a fraction of the acid will react to produce products. The acid dissociation or ionization constant, K_a, for the equation (1) can be written as:

$$K_a = \frac{[H_3O^+][B^-]}{[HB]} \qquad (2)$$

Where, the brackets indicates the equilibrium molar concentrations.

If we prepare an acid solution with an initial concentration of **(a)** moles/L, at equilibrium the concentration of acid, apparently will be less than **(a)** moles/L. Assuming that the reduction in the concentration of acid to be **(x)** moles/L, then, the concentrations of H_3O^+, and B^- both will be **(x)** moles/L and the equilibrium concentration of acid would be **(a-x)**. Replacing these concentration values in equation (2) we obtain:

$$K_a = \frac{x^2}{(a-x)} \qquad (3)$$

As stated above, $[H_3O^+]$ = **(x)** ; and **(a)** is the initial concentration of acid. One way to determine $[H_3O^+]$ is to use a pH-meter. The pH-meter directly measures the pH of a solution .
The relationship between pH and $[H_3O^+]$ is given as:

$$pH = - \log[H_3O^+] \qquad (4)$$

By rearranging equation (4) we can calculate $[H_3O^+]$ as:

$$[H_3O^+] = 10^{-pH} \qquad (5)$$

Therefore, the concentration of H_3O^+ in a solution could be determined by using the measured pH of the solution in equation (5).

To determine the value of K_a for a weak acid, we can use equation (3) for an aqueous solution of the acid with a known concentration of **(a)** moles/L. The only measurement required will be the determination of the pH of solution using a pH-meter. This process could be done for various concentrations of the same acid in order to obtain an average value for K_a. However, we are going to use another method to determine K_a for a weak acid.

A solution containing both a weak acid and its conjugate base is referred as a buffer solution. We can make a buffer by adding a strong base to the solution of a weak acid. A strong base in an aqueous solution can be represented by OH^-. The reaction of a strong base with a weak acid is:

$$HB(aq) + OH^-(aq) \text{ -----> } H_2O + B^-(aq) \qquad (6)$$

Unlike the reaction of a weak acid with water, this reaction goes to completion. Therefore, if initially, the number of moles of OH^- is less than the number of moles of acid, all the base will react with acid. At the end of the reaction, the number of moles acid will equal the initial number of moles of acid less the number of moles of base added. The number of moles of B^- will equal the number of moles of base added. The concentration of each species will equal the number of moles of that species divided by the TOTAL volume of the solution. So,the expression for K_a can be written as:

$$K_a = \frac{[H_3O^+][B^-]}{(\mathbf{a} - [B^-])} \qquad (7) \qquad [H_3O^+] = \frac{K_a\ (\mathbf{a} - [B^-])}{[B^-]}$$

$$\frac{1}{[B^-]} = \frac{[H_3O^+]}{\mathbf{a}K_a} + \frac{1}{\mathbf{a}}$$

This is the equation of a straight line. We can plot $1/[B^-]$ (y axis) versus $[H_3O^+]$ (x axis) and determine the slope of the line which is equal to $1/\mathbf{a}K_a$. Therefore, K_a is equal to (1/(a x slope) of the line.

The magnitude of K_a gives us a means of determining the extent to which an acid ionizes. Another means of ascertaining this is to determine the percent of dissociation or ionization. The percent ionization gives us the ratio of ionized molecules based on percent. Mathematically it is :

$$\%\text{Ionization} = \frac{[H_3O^+] \text{ from HA}}{[HA] \text{ original acid concentration}} \times 100$$

EXPERIMENTAL PROCEDURE

1. Turn the pH-meter on while you are preparing the solutions.

2. Wash and clean two burets, one for the standard unknown acid and the other for standard NaOH solution. Rinse each buret twice with a 5 mL portion of standard solution before filling it up.

3. Obtain 200mL of 0.600 M acid in a dry and clean flask, and 140 mL of 0.300M NaOH solution in another dry and clean flask.

4. Fill the acid rinsed buret with acid and the base rinsed buret with base. Set the meniscus bottom at the zero reading on each buret. Make sure that the tip of each buret is filled with liquid and that no air remains in it.

5. Prepare the following four solutions in clean 100mL volumetric flasks:

Solution number	0.6 M acid (mL)	0.3 M base (mL)	Volume Total* (mL)
1	50	20	100
2	50	30	100
3	50	40	100
4	50	50	100

*Dilute each to 100mL with distilled water if needed.

6. The pH-meter should be standardized with one or two standard buffer solutions. Ask your instructor for the procedure.

7. Mix the solutions in each flask.

8. Clean a 100 mL beaker and rinse it with about 5 mL of solution number 1. Add enough of solution number one in the beaker and measure the pH.

9. Repeat the same measurement for solutions 2-4.

10. Calculate **(a)**, and B^- concentrations for each solution

11. Plot $1/[B^-]$ vs $[H_3O^+]$ the data to obtain the slope and K_a.

Precaution 1. NaOH is toxic and corrosive.

Calculations:

1. The initial number of moles of acid in each solution is:
(50mL acid)x(0.600 mol/1000 mL) **or** (0.05 L x 0.6mol/L) = moles acid initially

2. The number of moles of B^- is equal with the number of moles of added base
(V mL of base)x(0.300 mol/1000mL) **or** (V base (liters) x 0.3mol/liter) = moles base = B^-

3. The total volume of each solution is 100 mL = 0.100 L.

4. The concentrations of **(a)** and B^- = mole of each / Volume Total in L.

EXPERIMENT 19 IONIZATION CONSTANT OF A WEAK ACID, K_a

DATA AND LAB REPORT SHEET

NAME__________________________ ASSIGNED #_____ DATE______ SCORE______

UNKNOWN #_____

SOLUTIONS

	1	2	3	4
Unknown acid				
Final buret reading, mL	______	______	______	______
Initial buret reading,mL	______	______	______	______
Volume of acid, mL	______	______	______	______
Concentration of acid =(Vx M initial)/V total	______	______	______	______
NaOH solution				
Final buret reading, mL	______	______	______	______
Initial buret reading,mL	______	______	______	______
Volume of NaOH, mL	______	______	______	______
Concentration of NaOH =(Vx M initial)/V total	______	______	______	______
Moles of conjugate base	______	______	______	______
$[B^-]$	______	______	______	______
$1/[B^-]$	______	______	______	______
pH of the Solutions	______	______	______	______
$[H_3O^+]$	______	______	______	______

Slope of Plot ____________

K_a from the slope, $K_a = \dfrac{1}{a \times \text{slope}}$ ____________

% Ionization of acid= $[[H_3O^+]/[HB]] \times 100$

______ ______ ______ ______

Average % Ionization of acid ______________

Show your calculations on lineless paper.

Pre-Laboratory Assignment Experiment 19
Ionization Constant of a Weak Acid, K_a
Name______________________________#______Score______

1. A 0.100 M solution of acetic acid is prepared. Calculate the pH of this solution.
   $K_a = 1.8 \times 10^{-5}$

2. A 0.50 M solution of an unknown acid has a pH = 2.20. Calculate the K_a of this acid.

3. A solution is prepared by adding 0.100 mol of lactic acid and 0.0200 mol of sodium hydroxide in 250 mL. The K_a for the acid is 1.4×10^{-4}, calculate the pH of this solution.

4. What is the pH of a solution made by adding 0.100mol of HCl, a strong acid, and 0.0800 mol of NaOH in enough water to make 500 mL of solution?

10 DIVISIONS PER CENTIMETER

EXPERIMENT 20

SOLUBILITY PRODUCT CONSTANT -Ksp

OBJECTIVES

a) To determine a solubility product constant
b) To calculate the molar solubility of a salt
c) To understand the relationship between molar solubility and K_{sp}
d) To become familiar with equilibria involving a sparingly soluble salt

SUPPLIES AND EQUIPMENT

Spectronic 20, Burets or pipets, 2E-3M $Pb(NO_3)_2$, 0.25M KNO_3, 2E-3M K_2CrO_4, Cuvets, Distilled Water, Kimwipes, Centrifuges, Graph Paper, 100mL Volumetric Flasks

WHAT YOU SHOULD KNOW BEFORE YOU BEGIN

In this experiment, you will determine the solubility product constant of Lead(II) chromate, $PbCrO_4$. The equation of the equilibria is:

$$PbCrO_4(s) \Longleftrightarrow Pb^{+2}(aq) + CrO_4^{-2}(aq)$$

If the chromate ion is added to a saturated solution of lead chromate solution, lead chromate will precipitate due to the shift in the equilibrium as shown above. Lead chromate is a sparingly soluble salt. Salts such as the $NaNO_3$, KNO_3, or KCl are readily soluble and are strong electrolytes. In-order for equilibrium to be established between a sparingly soluble salt and its ions, the solution must be saturated. The equilibrium constant expression for the reaction is:

$K_{sp} = [Pb^{+2}][CrO_4^{-2}]$, the product of the ion concentrations to the coefficient powers in moles / liter. Using the formula of the salt, the molar solubility is the concentration of the Pb^{+2} ion, or that of the CrO_4^{-2} ion because it is a 1:1 salt. The molar solubility is designated as (s). To find the molar solubility from K_{sp}, the expression used would be: $K_{sp} = (s)(s) = s^2$, the square root would be taken.
In a salt such as PbI_2, the molar solubility of which is equal to the concentration of Pb^{+2} ion which is (s), while the concentration of I^- would be (2s).

$$\underset{}{PbI_2(s)} \Longleftrightarrow \underset{(s)}{Pb^{+2}(aq)} + \underset{(2s)}{2I^{-1}(aq)}$$

The solubility product constant expression would be : $K_{sp} = [Pb^{+2}][I^-]^2$. When finding the molar solubility from the K_{sp} value, the expression becomes:
$K_{sp} = (s)(2s)^2 = 4s^3$ the cube root would be taken. In the equilibria under study here, the chromate ion will be analyzed spectrophotometrically using a Spectronic 20. The chromate ion is bright yellow in aqueous medium. The absorption maximum occurs at 375nm (3750Å). Therefore, you will use this value of wavelength for the analyses.

A standard curve will be prepared of known [CrO_4^{-2}] solutions, and the absorbance of each solution determined. A plot will be made of absorbance versus concentration of the CrO_4^{-2} ion. A set of chromate solutions will be prepared for analyses and subsequent K_{sp} and molar solubilities determined. In regards to precipitation, a solid will precipitate from an aqueous solution of its ions when the ion product concentrations is just greater than the known K_{sp}. If the ion product concentrations, Q = the K_{sp} the reaction is at equilibrium.
If Q is $> K_{sp}$, then precipitation will occur and the reaction shifts to the left where solid is formed. $PbCrO_4(s) <===> Pb^{+2}(aq) + CrO_4^{-2}(aq)$

<-----------

If the ion product concentrations, Q, is $< K_{sp}$, the ions are present in solution only, and the reaction goes to the forward direction (right).

$PbCrO_4(s) <===> Pb^{+2}(aq) + CrO_4^{-2}(aq)$

------------>

The solubility product constant is the equilibrium constant of a slightly soluble ionic compound.

Given the following for PbI_2, $K_{sp} = 7.1E\text{-}9$. Will the salt precipitate from an aqueous solution that is prepared by adding 5mL of 1.0E-4M $Pb(NO_3)_2$ to 5mL of 1.2E-3M NaI?

$PbI_2(s) <===> Pb^{+2}(aq) + 2I^{-1}(aq)$

$K_{sp} = [Pb^{+2}][I^{-1}]^2 = 7.1E\text{-}9$ Precipitation should occur if $[Pb^{+2}][I^{-1}]^2 > 7.1E\text{-}9$.

$[Pb^{+2}]$ = V X M / V Total = 5E-3liters x 1E-4M / 0.01liters = 5E-5M

$[I^{-1}]$ = V X M / V Total = 5E-3liters x 1.2E-3M / 0.01liters = 6E-4M

$[Pb^{+2}][I^{-1}]^2$ = 5E-5M x $(6E\text{-}4M)^2$ = 1.8E-11

Since 1.8E-11 is $<$ 7.1E-9, the salt will **not precipitate** from an aqueous solution.

If 5mL of 1.5E-2M $Pb(NO_3)_2$ was added to 5mL of 1.2E-2M NaI, a precipitate will form:

$[Pb^{+2}]$ = V X M / V Total = 5E-3liters x 1.5E-2M / 0.01liters = 7.5E-3M

$[I^{-1}]$ = V X M / V Total = 5E-3liters x 1.2E-2M / 0.01liters = 6E-3M

$[Pb^{+2}][I^{-1}]^2$ = 7.5E-3M x $(6E\text{-}3M)^2$ = 2.7E-7

Since 2.7E-7 is $>$ 7.1E-9, the salt **will precipitate** from an aqueous solution.

EXPERIMENTAL PROCEDURE
(Designed for groups)

A. STANDARD CURVE

1. Obtain a clean buret, rinse it with distilled water and with 2E-3M K_2CrO_4, and fill it with 2E-3M K_2CrO_4, alternatively you may use a pipet.

2. Place approximately 500mL of 0.25M KNO_3 in a 600mL beaker to be used for dilution to the calibration mark in each volumetric flask.

3. Obtain five 100mL Volumetric flasks and label from #1-#5.
   Dispense from the buret the specified quantities as indicated in Table 20.1 below and dilute each to the 100mL calibration mark with 0.25M KNO_3.

Table 20.1. Standard Mixtures K_{sp}

Flask #	K_2CrO_4
1	0.5mL
2	1.5mL
3	3.5mL
4	5.5mL
5	7.5mL

*dilute to 100mL volume with 0.25M KNO_3
The KNO_3 is used as the blank in the analyses.

4. Pour appropriately each sample into a cuvet, measure and record the absorbance/ transmittance of each sample using the Spectrophotometer, the use of which will be explained by the laboratory instructor.

5. Plot on linear graph paper the Absorbance vs [CrO_4^{-2}] and draw the best straight line through the data points.

B. EQUILIBRIUM UNKNOWN SOLUTIONS

1. Obtain a clean buret,- rinse it with distilled water and with 2E-3M $Pb(NO_3)_2$, and fill it with $Pb(NO_3)_2$.

2. Obtain four test tubes and label from #1-#4.
   Buret or pipet in the specified quantities as indicated in Table 20.2.

Table 20.2. Equilibrium Unknown Solutions

Test tube #	K_2CrO_4	$Pb(NO_3)_2$
1	5mL	5mL
2	5mL	5mL
3	5mL	5mL
4	5mL	5mL

3. Vigorously shake the test tubes for a minimum of four minutes. Save

4. Transfer 2 ml of each solution into a centrifuge tube and centrifuge each sample and discard the supernatant (liquid) from each . Save the precipitates.

5. Wash each precipitate with KNO_3 by adding approximately 1 ml of 0.25M KNO_3 and shaking the test tube, followed by centrifuging and discarding the supernatant.

6. Add 3mL of 0.25M KNO_3 to each tube and shake for an additional 20 minutes to allow the establishment of the equilibrium system. However, if there is no solid remaining, use the remainder of the reaction mixture from step 2 . Centrifuge the samples.

7. Pour the clear supernatants into cuvets for absorbance measurements. Measure and record the absorbance of each sample. Using the standard plot, determine and record the value of the [CrO_4^{-2}] for each unknown.

Precautions: 1. The blank solution contains the solvent system utilized in the experiment.
2. Always rinse the cuvet with the next sample to be analyzed or use a clean dry cuvet.
3. Always wipe the cuvet to remove all foreign substances including finger prints before insertion into the Spectrophotometer
4. The chromate solutions should be poured into the waste containers.

Sample K_{sp} Calculations

e.g. The absorbance of unknown 1 solution was found to be 0.2

Using the standard curve the [CrO_4^{-2}] was found equal to 1.2E-7M

[Pb^{+2}] = [CrO_4^{-2}] from the balanced equation

The K_{sp} = [Pb^{+2}][CrO_4^{-2}] = 1.2E-7 x 1.2E-7 = 1.44E-14

The Molar solubility = 1.2E-7M

or from K_{sp} , 1.44E-14 = (s)(s) = s^2, s= $(1.44E-14)^{1/2}$ = 1.2E-7

EXPERIMENT 20 SOLUBILITY PRODUCT CONSTANT-K_{sp}

DATA AND LAB REPORT SHEET

NAME________________________________ASSIGNED #_____ DATE______ SCORE______

A. STANDARD CURVE

$[CrO_4^{-2}]$ = ____2E-3M___________M

Standard Mixtures

	1	2	3	4	5	
Vol- CrO_4^{-2}	______	______	______	______	______	L
Moles-CrO_4^{-2}	______	______	______	______	______	mol
$[CrO_4^{-2}]$	______	______	______	______	______	M
Absorbance, A	______	______	______	______	______	

B. EQUILIBRIUM UNKNOWN SOLUTIONS

Unknown Mixtures

	1	2	3	4	5	
Absorbance, A - $[CrO_4^{-2}]$	______	______	______	______	______	
$[CrO_4^{-2}]$ from Calibration Curve	______	______	______	______	______	M
$[Pb^{+2}]$ from Stoichiometric Equation	______	______	______	______	______	M
K_{sp}	______	______	______	______	______	

K_{sp} average ____________________

K_{sp} standard deviation ____________________

What the molar solubility of $PbCrO_4$? ____________________

Show Determinations on lineless paper completely for any one set (column).

QUESTIONS

1. In a saturated PbI_2 solution, the molar concentration of I^{-1} measured spectrophotometrically was found to be 2.6E-3M. What is the Ksp of PbI_2 ?

2. When will a slightly soluble salt precipitate from an aqueous solution of its ions?

Pre-Laboratory Assignment Experiment 20
Solubility Product Constant Name________________________#________Score________

1. If the Ksp of $BaCrO_4$ is 1.1E-5, what is the molar solubility?

2. Write the reaction equilibrium for Ag_2CO_3. If the molar solubility of Ag_2CO_3 is 1.3E-4, what is the K_{sp}?

3. What is the color of the CrO_4^{-2} ion?

4. The Ksp of Ag_2CrO_4 is 9E-12. If 5mL of 4.0E-3M $AgNO_3$ is mixed with 5mL of 2.4E-3M K_2CrO_4, will a precipitate form? show determination

10 DIVISIONS PER CENTIMETER

EXPERIMENT 21

HEAT OF REACTION

OBJECTIVES

a) To determine the heat capacity of a coffee cup calorimeter
b) To determine the molar heat of neutralization of a strong acid-base system

SUPPLIES AND EQUIPMENT

1.5 M NaOH, 1.5 M HCl, Coffee Cups and Lids, Thermometer, 100mL Graduate Cylinder

WHAT YOU SHOULD KNOW BEFORE YOU BEGIN

Most chemical reactions involve some heat exchange. That means, chemical reactions either absorb or evolve some heat. If heat was absorbed in a process it is called an endothermic process. The process that evolves heat is said to be exothermic. An endothermic process will cause a cooling on the surrounding (the container, solution, etc.). The heat released in an exothermic process will warm the surrounding.

The amount of heat can be determined from the following formula, $q = (s)m(t_f - t_i)$ (1) where, **q** is the heat absorbed or evolved, **(s)** stands for the specific heat (the amount of heat required to raise the temperature of one gram of a substance by one degree), **(m)** is the mass of substance to be heated, t_f is the final temperature, and t_i is the initial temperature.

The amount of heat absorbed or evolved depends on a property of a substance called the enthalpy, H. The heat content of a substance is known as the enthalpy. For a chemical reaction an endothermic process means that the enthalpy of products are larger than that of reactants [H(P) - H(R) is a positive number]. Exothermic reactions means that the enthalpy of products are smaller than that of reactants [H(P) - H(R) is a negative number].

Enthalpy like density, color, boiling point melting point is a property of a substance and is an extensive property (depends on the amount of substance used). However, the absolute value of it has no meaning and cannot be determined. What we are interested is the change in enthalpy, ΔH. Enthalpy is a state property, that means it only depends to the initial and final state of matter and is independent of the path the change might take. State properties are very useful in determining physical quantities for if the conditions change during an experiment one could bring them back to the original condition (T, P, ..) and obtain the correct result.

When a process occurs at constant pressure the amount of heat liberated or absorbed is equal to ΔH. In the laboratory for an open system the pressure is constant during the experiment. Therefore, by measuring the amount of heat we will be able to determine ΔH of the reaction. The heat evolved by the reaction is absorbed by water and by the coffee cup calorimeter.
ΔH = heat absorbed by water + heat absorbed by the calorimeter
$\Delta H = -[q(\text{water}) + q(\text{calorimeter})]$ (2)

EXPERIMENTAL PROCEDURE

A. THE HEAT CAPACITY, s.m., OF THE CALORIMETER

The specific heat of water from 0 to 100 degree Celsius is about the same. Therefore, it is expected that if equal masses of water at two temperatures were added together and mixed the temperature of the mixed water to be the average value of the original temperatures. However, some heat will be absorbed by the calorimeter, assuming no heat is lost to the surrounding:

$$\Delta H = 0 = C(t_f - t_c) + (s)m_1(t_f - t_c) + (s)m_2(t_f - t_h) \qquad (3)$$

$(s) = 4.184 J/g^oC$ for water

Heat lost by the hot water = heat gained by the cold water + Heat gained by the calorimeter.

1. Pick a clean and dry coffee-cup and a clean and dry 250 mL beaker. Using a graduated cylinder, carefully transfer 50 mL of warm water (at about 50^oC) into the beaker and using clean and dry graduated cylinder transfer 50 mL of water (at room temperature) into the coffee-cup.

2. Measure the temperature of the water in the beaker and in the coffee-cup with a thermometer. If you are using one thermometer, please be sure to rinse and wipe clean before inserting from one solution to another.

3. Transfer the water in the beaker into the coffee-cup to the last drop as quickly as possible. While stirring with the thermometer determine the final temperature of the mixed water.

4. By replacing the numbers in equation (3) determine C, the heat capacity of calorimeter in $J/^oC$.

5. Repeat the above procedure for trial 2.

B. DETERMINATION OF THE HEAT OF REACTION

1. Place 50.0 mL of 1.5 M HCl in the coffee-cup, and 50.0 mL of 1.5 M NaOH solution in the beaker. Monitor the temperatures and be sure that the temperatures of the two solutions are not different by more than 0.5 ^oC.

2. Record the temperatures and transfer the solution in the beaker into the coffee-cup and record the temperature of the solution as you are stirring with the thermometer every 30 seconds. Obtain the highest temperature .

3. Repeat the above procedure for trial 2

Calculations:

$q(w) = [C + (s)m](t_f - t_i)$, (s) for the solution can be taken as that of pure water. The density of the solution is close to 1.0 g/mL ,

$m = d.v =$ 100mL of solution x 1.0g/mL = 100g,

n (moles) = Volume(L) x Molarity = L.M, $\qquad \Delta H = -q/n$

EXPERIMENT 21 HEAT OF REACTION

DATA AND LAB REPORT SHEET

NAME________________________________ASSIGNED #_____ DATE______ SCORE______

A. THE HEAT CAPACITY, s.m., OF THE CALORIMETER

Trial 1		Trial 2	
Temp.^oC of warm water	Temp.^oC of cold water	Temp.^oC of warm water	Temp.^oC of cold water
________	________	________	________
________	________	________	________

Trial 1	Trial 2
Temp.^oC of the mixed water (30 second intervals)	Temp.^oC of the mixed water (30 second intervals)
____________________	____________________
____________________	____________________
____________________	____________________
____________________	____________________
____________________	____________________
____________________	____________________
____________________	____________________
____________________	____________________
____________________	____________________
____________________	____________________
Final temperature __________	Final temperature __________

(Show calculation for one trial)	Trial 1	Trial 2
The heat lost by the hot water	______________	______________
The heat gained by the cold water	______________	______________
The heat gained by the coffee-cup	______________	______________
Heat capacity of the coffee-cup J/^oC	______________	______________

Average Heat capacity of the coffee-cup J/^oC ________________

B. DETERMINATION OF THE HEAT OF REACTION(ACID-BASE)

(Show calculation for one trial)

Trial 1		Trial 2	
T°C HCl	T°C NaOH	T°C HCl	T°C NaOH
________		________	
________		________	
________		________	
________		________	
________		________	
________		________	
________		________	

Final Temperature________________

	Trial 1	Trial 2
Moles of acid reacted	________	________
Moles of salt produced	________	________
Moles of water produced from the reaction	________	________
The heat gained by the solution	________	________
The heat gained by the coffee-cup	________	________
ΔH of the reaction for the amounts used	________	________
Molar enthalpy of the reaction	________	________

Average Molar enthalpy of the reaction ________________

Pre-Laboratory Assignment Experiment 21
Heat of Reaction Name______________________________#______Score______

1. 30.0 mL of water at 24.5°C was mixed with 40.0 mL of water at 40°C. Taking the specific heat of water as 4.184 J/g°C, and assuming that no heat was absorbed by the calorimeter calculate the final temperature.

2. If the final temperature in question (1) was 33.1°C, calculate

   (a) the amount of heat, in joules, lost by warm water

   (b) the amount of heat gained by the cold water

   (c) the heat capacity of the calorimeter

3. In a coffee-cup calorimeter the following reaction is carried out: $H^+(aq) + OH^-(aq) \text{---}> H_2O(l)$ When 0.015 mol of H^+ reacts with 0.015 mol of OH^-, the temperature of 110 g of water rises from 24.0 to 25.8°C. Ignoring the heat absorbed by the coffee-cup calculate

   (a) q(water)

   (b) ΔH for the reaction

   (c) ΔH when 1.00 mol of H^+ reacts with 1.00 mol of OH^-.

4. If we use a weak acid instead of hydrochloric acid, do you think that the ΔH will be larger or smaller? explain.

EXPERIMENT 22
QUALITATIVE ANALYSIS

OBJECTIVES

a) To apply the concepts of pH, Ksp, redox reaction, and Le Chatliers principle to Qualitative Analysis.
b) To be able to separate and identify common cations in Qualitative Analysis.
c) To be able to write net ionic equations for all the displacement reactions in Qualitative Analysis.
d) To be able to critically think through a series of experimental observations and predict the specific products.

SUPPLIES AND EQUIPMENT

Centrifuge, Test Tubes, Droppers, Litmus Paper, Stirring Rods, Centrifuge Tubes, Beakers (150 mL, 250 mL), Bunsen Burner, Lighter, Test Tube Holder
Reagents:
Unknown Solution

6 M HCl, 1 M HCl, 18 M H_2SO_4, 6 M H_2SO_4, 6 M HNO_3,
6 M CH_3COOH

15 M NH_3, 6 M NH_3, 3 M NH_3, 1 M NH_3, 4 M NaOH

0.5 M K_2CrO_4, 1 M NH_4I, 0.2 M $HgCl_2$, 1 M NH_4NO_3,
1 M NH_4CH_3COO, 1 M $SnCl_2$, 0.3 M $K_4Fe(CN)_6$, 3 M DMG,
4 M NH_4SCN, 1 M $NaNO_2$, , 0.1 M $Pb(CH_3COO)_2$

5 % Thioacetamide, 6% H_2O_2, 5% NaClO, 5% $(NH_4)_2S$

Solid $Na_2S_2O_4$, Solid $H_2C_2O_4.2H_2O$, Solid $NaBiO_3$,

WHAT YOU SHOULD KNOW BEFORE YOU BEGIN

Qualitative Analysis is an integral part of the general chemistry course. Here you will observe accurately and interpret the results logically. This is one of the most important skills in the learning process. As the name suggests, Qualitative Analysis involves the identification of ions in a solution. The analysis is based on some important concepts in chemistry, including pH, redox reactions, Ksp, and Le Chatliers principle, which you should be familiar with before beginning qualitative analysis. As you will proceed, these are the concepts you will be applying to interpret your observations.

Preview the concepts in the text before beginning of this laboratory.

During the next few weeks you will be involved in the identification of cations in an aqueous solution. Initially you will selectively precipitate the ions in groups as insoluble compounds from a starting solution. You will then separate these ions and identify each one based on their chemistry.

GROUP I CATIONS:

Silver (Ag^{+}), mercury (Hg_2^{2+}), and lead (Pb^{2+}) ions are the three cations in Group I. When hydrochloric acid is added to the solution with all other common cations these three cations will precipitate out as chloride salts. These are separated from the solution by centrifugation and filtration.

GROUP II CATIONS:

After removal of Group I cations, when the solution is made 0.3 M in HCl, and is saturated with hydrogen sulfide, Group II cations will precipitate out as sulfides. These ions are lead (Pb^{2+}), bismuth (Bi^{3+}), copper (Cu^{2+}), cadmium (Cd^{2+}), mercury (Hg_2^{2+}), arsenic (As^{3+}), antimony (Sb^{3+}), and tin (Sn^{4+}) sulfides.

GROUP III CATIONS:

After the Group II cations have been removed, to precipitate the group III cations, the solution is saturated with hydrogen sulfide, and is made basic with aqueous ammonia (slight excess). The sulfides of cobalt (Co^{2+}), nickel (Ni^{2+}), manganese (Mn^{2+}), iron (Fe^{2+}), and zinc (Zn^{2+}), together with aluminum (Al^{3+}) and chromium (Cr^{3+}) hydroxides precipitate out of the buffered basic solution.

TECHNIQUES:

Several techniques will be used frequently in the qualitative analysis scheme. Some of these are described below:

Burner:

Be careful to light up a burner and use the flame properly. When heating is desired, you may have to start slowly for a gentle heating to digesting a solution over a fixed length of time. You may have to boil down the solution to a certain volume or gently warm a solution to a certain degree. Be sure to use a flame of proper height and for an appropriate length of time.

Stirring Rods:

Also see Cutting of Glass Technique 23T on Page 18

You will require several stirring rods for mixing purposes. Make several of these from 3 mm diameter glass rod. Each rod can be between 15 to 30 cm long. For this ay the rod on the bench top, and then make a scratch by using the file in your drawer. Now placing the scratch away from you push the rod out with your thumb behind the scratch. Then fire polish the ends.

Heating a Solution:

Also see Heating a solution, Technique 8T on Page 10

A solution can be easily heated by placing it in the test tube and then putting the test tube in a hot water bath. Do so unless you are told otherwise. This is a safe way to heat solutions in qualitative analysis.

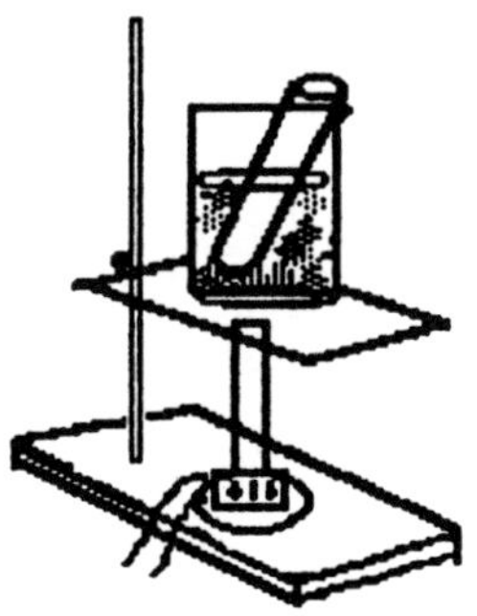

Capillary Pipets:

For withdrawing solutions from a test tube, it is convenient to use a capillary pipet rather than a medicine dropper. You can make several capillary pipets from 8 mm glass tubing. Cut a piece of 20 cm tubing, evenly heat it in the Bunsen burner flame by continuously rotating it, and then finally pull the two ends of the tube apart while removing it from the flame. Let the glass cool and then cut the glass where the tube is thin. Fire polish the wider end by rotating it in the flame until it glows red for 30 seconds and then letting it cool down. Attach rubber bulbs at the end before use.

Evaporating Dish:

Several times during the analysis you will be required to reduce the volume of a solution by evaporation. Sometimes it is convenient to use an evaporating dish for this purpose. Place the evaporating dish with the solution on a clay triangle. Start heating the solution by slowly passing a low flame from side to side under the dish. Be careful when the volume is too low as spattering will occur which will result in a loss of your ions or compounds.

Centrifugation and Separation of a Precipitate:

During the course of experiments, sometimes a solid precipitate will result when you mix different reagents. You will be required to separate the precipitate by centrifugation. In order to do so your mixture must be placed in a small test tube. Take another test tube of the same size and fill it with an equal volume of water. Now place both the test tubes in a centrifuge (instructor will show you) <u>**across from each other**</u> and switch it on. After about 2 minutes, switch off the centrifuge. Allow the cetrifuge to come to a **complete stop**. Remove the test tubes and you will observe solid at the bottom of the test tube with the supernatant liquid at the top. You can use a capillary pipet to remove most of the liquid at the top.

Washing a Precipitate:
After centrifuging a suspension, the solids will separate from the liquid and will settle down at the bottom. You will be able to remove this decantate (supernatant) by using a capillary pipet. On several occassions, the precipitate needs to be washed to remove the residual liquid, which may interfere with the analysis. Therefore, to wash the precipitate, add the wash liquid (as instructed) and stir well. Centrifuge and separate the liquid. Follow the instructions to either save the liquid with the previously removed decantate, or to discard it.

Mixing Reagents:
It is very important that reagents are mixed well for a reaction to be completed in a certain period of time. To mix the reagents, after you have placed them in a test tube, use a **clean** stirring rod and mix well.

Testing the Acidity/Basicity of a Solution:
The simple way to test whether a solution is acidic or not is by using blue litmus paper which turns red when the solution is acidic. To test, put a glass rod in the solution and remove it. Now touch the end of the rod to a piece of blue litmus paper on a watch glass. If it turns red then the solution is acidic. Use the dry portion of the litmus paper each time you test the solution. The same procedure can be followed to test whether a solution is basic or not is by using red litmus paper which turns blue upon contact with a basic solution.

The Procedure:
The analytical procedure must be understood well to smoothly carry out the tests. Several things you may want to observe in the instructions. If a precipitate is formed in the analysis, it is placed in boxes with a double border. The solutions (supernatant) are in boxes with single borders. A caution is issued with an ! sign. When a cation is confirmed it is underlined. Following the procedures correctly will lead to correct results. Otherwise you may have to start all over again.

ALERT! Be careful what you throw out!

Dispose of chemical wastes as directed by the instructor.

To give you a quick overview of qualitative analysis let us start with a flow chart for the separation of some common cations.

Recognize that a group number in the qualitative scheme does not refer to a group in the periodic table.

Ag^+ Pb^{2+} Hg_2^{2+}	Pb^{2+} Bi^{3+} Cu^{2+} Cd^{2+} Hg^{2+} $[SbCl_4]^-$ $[SnCl6]^{2-}$ H_3AsO_4 $[HgCl_4]^{2-}$	Co^{2+} Ni^{2+} Fe^{3+} (Fe^{2+}) Mn^{2+} Al^{3+} Cr^{3+} Zn^{2+}	Ba^{2+} Sr^{2+} Ca^{2+}	Mg^{2+} K^+ Na^+ NH_4^+

↓ 6 M HCl

Group I - The Insoluble Chloride Group

$AgCl$ $PbCl_2$ Hg_2Cl_2	$PbCl_2$ Bi^{3+} Cu^{2+} Cd^{2+} Hg^{2+} $[SbCl_4]^-$ $[SnCl6]^{2-}$ H_3AsO_4 $[HgCl_4]^{2-}$	Co^{2+} Ni^{2+} Fe^{3+} (Fe^{2+}) Mn^{2+} Al^{3+} Cr^{3+} Zn^{2+}	Ba^{2+} Sr^{2+} Ca^{2+}	Mg^{2+} K^+ Na^+ NH_4^+

↓ HNO_3 evaporate to moist residue

$HSb(OH)_6$ $Hsn(OH)_6$	Pb^{2+} Bi^{3+} Cu^{2+} Cd^{2+} Hg^{2+} H_3AsO_4	Co^{2+} Ni^{2+} Fe^{3+} Mn^{2+} Al^{3+} Cr^{3+} Zn^{2+}	Ba^{2+} Sr^{2+} Ca^{2+}	Mg^{2+} K^+ Na^+ NH_4^+

H_2S ↓ 0.3 M HCl, I^- (catalyst)

Group II - The Insoluble (acidic medium) Sulfide Group

PbS Bi_2S_3 CuS CdS Sb_2S_3 SnS_2 As_2S_3 HgS	Co^{2+} Ni^{2+} Fe^{3+} Mn^{2+} Al^{3+} Cr^{3+} Zn^{2+}	Ba^{2+} Sr^{2+} Ca^{2+}	Mg^{2+} K^+ Na^+ NH_4^+

$(NH_4)_2S$ ↓ aq NH_3, NH_4Cl

Group III- The Insoluble (basic medium) Sulfide Group

CoS NiS FeS MnS $Al(OH)_3$ $Cr(OH)_3$ ZnS	Ba^{2+} Sr^{2+} Ca^{2+}	Mg^{2+} K^+ Na^+ NH_4^+

QUALITATIVE ANALYSIS PRECIPITATES AND AQUEOUS SOLUTIONS OF IONS

SOLID	COLOR	SOLID	COLOR	SOLID	COLOR
$AgCl$	White	$PbCl_2$	White	Hg_2Cl_2	White
$PbCrO_4$	Yellow	CuS	Black	CdS	Yellow
$HgNH_2Cl$	White	$Hg(l)$	Black	$Al(OH)_3$	White
Bi_2S_3	Brown-Black	Sb_2S_3	Orange-red	MnO_2	Brown-Black
As_2S_3	Yellow	PbS	Black	$Ni(DMG)_2$	Red
HgS	Black	SnS_2	Yellow	FeS	Black
$Bi(OH)_3$	White	$Cd(OH)_2$	White	CoS	Black
$Cu_2Fe(CN)_6$	Red-Brown	$Cd_2Fe(CN)_6$	Yellow-White	ZnS	White
$Cu(OH)_2$	Blue-green	$MgNH_4AsO_4$	White	MnS	Faint pink
Ag_3AsO_4	Red-Brown	$Hg(l) + Hg_2Cl_2$	Gray	NiS	Black
Bi	Black	$PbSO_4$	White	$Cr(OH)_3$	Green
$Al(OH)_3$+ aluminon	Red gel like lake				

SOLUTIONS OF IONS

ION	COLOR	ION	COLOR	ION	COLOR
$Cu(H_2O)_6^{+2}$	Aqua	$Cu(NH_3)_4^{+2}$	Deep Blue-violet	Fe^{+2}	Pale Green
$FeSCN^{+2}$	Red	$Co(SCN)_4^{-2}$	Blue	$Cr_2O_7^{-2}$	Orange
$Cr(H_2O)_6^{+3}$	Purple	$Fe(H_2O)_6^{+3}$	Faint purple	CrO_4^{-2}	Yellow
$Mn(H_2O)_6^{+2}$	Faint Pink	$Ni(H_2O)_6^{+2}$	Green	MnO_4^{-1}	Purple

FLOW CHART for GROUP I ANALYSIS

UNKNOWN SOLUTION
Pb^{2+} Ag^{+} Hg_2^{2+}

1.Take about 1 mL of the solution to be analyzed in a test tube
2. Add 2 drops of **6 M HCl** to it , warm it in a hot water bath while stirring.
3. Cool the solution by placing the test tube in running water and centrifuge.*
4. *Test for complete precipitation with HCl. (**Refer to pre lab**)*
5. Separate the solution from the precipitate.
6. *If the unknown is a general unknown then save the solution otherwise discard it.*

i. Lead Chloride and Silver Chlorides may dissolve if the solution is too acidic. Therefore add just enough HCl for complete precipitation.

* cold water is used because $PbCl_2$ is soluble in hot water.

Answer Question 1-5 after step 6

Group II, III, IV,V

$PbCl_{2(s)}$ (White), AgCl $_{(s)}$(White), $Hg_2Cl_{2(s)}$ (White)

7. Add 1 mL of water to the precipitate, stir, and heat the test tube (***Refer to technique - Heating a test tube***).
8. Quickly centrifuge and remove the liquid to and save it in a test tube.
9. If precipitate remains then repeat steps 7 & 8 and save the liquid in the same test tube.

Answer questions 6-7 after step 9

AgCl $_{(s)}$(White), $Hg_2Cl_{2(s)}$ (White)

$PbCl_{2(aq)}$

10. Equally divide the solution into two test tubes

11. To one portion add 1 drop of ***conc.*** H_2SO_4. A white precipitate confirms Pb^{2+}

12. Add 5 drops of 0.5 M K_2CrO_4 to the second test tube A yellow precipitate confirms Pb^{2+}

! Be Careful with concentrated sulfuric acid, H_2SO_4.

Answer questions 8-9 after step 12

13. Add 0.5 mL of 6 M $NH_3(aq)$ to the precipitate, stir, and centrifuge. If a black precipitate forms separate the liquid and save. Repeat this step .

$Hg_{(s)}$ (black), $NH_2HgCl_{(s)}$(white). Confirms Hg_2^{2+}

$Ag(NH_3)_2Cl_{(aq)}$

14. Add 4 drops of 6 M HNO_3 at a time. Stir the mixture and test until the solution is acidic. A white precipitate confirms Ag^{+}.

Answer question 10 after step 14

ALL CHEMICAL WASTES MUST BE PLACED IN DESIGNATED CONTAINERS FOR PROPER DISPOSAL.

EXPERIMENT 22 QUALITATIVE ANALYSIS - GROUP I
DATA AND LABORATORY REPORT SHEET

Name ______________________ Assigned #____ Date _________ Score_______
Unknown #__________

1. Ions Present in the unknown (Group I) are ______________________________
2. The precipitating reagent for group I is
   a. Hot water
   b. Cold water
   c. 6 M HCl
   d. None is required

3. When an unknown solution is treated with HCl, if a precipitate forms, what ions are possible?

4. Is it necessary that to get a precipitate all of the ions in answer to the question 3 be present in the solution?

5. After finishing steps 1-7, what would be your reason to either save or discard the decantate?

6. Write the net ionic equation for the precipitation of all the ions in the answer to question 3?

7. In the step 8 why must you quickly centrifuge?

8. At step 9 if no precipitate remains, what would you infer?

9. If $PbSO_4$ does not form in step 11, but $PbCrO_4$ forms in step 12 , how will you explain these results? (Hint - Ksp)

10. Write the net ionic equation for the precipitation of $PbCrO_4$.

11. When $NH_3(aq)$ is added to AgCl (s), what is the physical change that you would observe?

Pre-Laboratory Assignment Experiment 22
Qualitative Analysis - Group I
Name ______________________________ #________ Score_____

1. How do you test for complete precipitation?

2. Why are Ag^+, Pb^{2+}, and Hg_2^{2+} grouped as Group I cations in the qualitative analysis scheme?

3. In the precipitation of Group I ions a little excess HCl is used. Why? Also why must you avoid using a large excess of HCl?

4. Given a pair of compounds, match with a reagent that will separate them.

a. $AgCl$, $PbCl_2$ (____)	i. HCl
b. Hg_2Cl_2, $PbCl_2$ (____)	ii. $NH_3(aq)$
c. $AgNO_3$, $Cu(NO_3)_2$ (____)	iii. Hot water
d. Hg_2Cl_2, $AgCl$ (____)	iv. HNO_3

5. $Ag^{+1} + 2NH_3 \Longleftrightarrow Ag(NH_3)_2^{+1}$
What is the concentration of Ag^{+1} in a 0.0023 M $[Ag(NH_3)_2]Cl$ and 0.680 M in NH_3 (aq) ? K_f = 1.7E7

6. A solution containing some unknown cations is treated with 6 M HCl. A white precipitate (A) was formed. When hot water was added to (A), it dissolved. To the resulting solution, 5 drops of K_2CrO_4 was added and a yellow precipitate (B) was formed. Identify (A) and (B).

FLOW CHART for GROUP II ANALYSIS

! Start this group carefully. A simple mistake like discarding a solution or a precipitate might require you to start over again. If such a situation arises consult the instructor.

* Use fresh known or unknown solution or if you are doing general unknown use the centrifugate from Group I separation

Pb^{2+} Bi^{3+} Cu^{2+} Cd^{2+} $[HgCl_4]^{2-}$ H_3AsO_4 $[SbCl_4]^-$ $SnCl_6]^{2-}$ *

A1. Add 4 drops of 6 M HNO_3 to 10 drops of the above solution and evaporate to a moist residue.(!).
A2. Cool the moist residue and add 10 drops of water & mix well.
A3. Add 1 M $NH_{3(aq)}$ until the solution is just basic.
A4. Add 1 M HCl until the solution is **JUST** acidic, and then add two drops of the 6 M HCl and a drop of 1 M NH_4I. Dilute the solution to 1.5 mL. Mix well.
A5. Heat the solution for 3 minutes. Add 12 drops of 5% thioacetamide solution. Heat the solution for 10 minutes, then add 1 mL of water and heat the solution for 5 more minutes. Stirr the solution often while stirring.
A6. Centrifuge and separate the precipitate. (!)
A7. If you are analyzing Genral Unknown, See NOTE 1

! (A1) Avoid evaporating the precipitate to dryness

Note 1: For general unknown only
a) Save the decantate from this step (6) for analysi of group III - V. Add 2 drops of 12 M HCl to the centrifugarte (the solution) and place it in hot water bath.
b) Wash the precipitate twice with 10 drops of 0.1 M HCl. Save the first 10 drops with the original centifugate. Discard the second 10 drops.

PbS Bi_2S_3 CuS CdS HgS As_2S_3 Sb_2S_3 SnS_2

B1. Add 12 drops of 4M NaOH and 5 drops of thioacetamide solution.
B2. Heat the solution in a water bath for 5 minutes, stirring constantly.
B3. Centrifuge and separate the solution. **Save the solution in a test tube lebeled *GR2B*.** This will be treated according to procedures in box F

PbS Bi_2S_3 CuS CdS HgS

$[HgS_2]^{2-}$ $[As(OH)_4]^-$ $[AsS_3]^{3-}$ $[Sb(OH)_4]^-$ $[SbS_3]^{3-}$ $[Sn(OH)_6]^{2-}$ $[SnS_3]^{2-}$

C1. Wash the precipitate with a mixture of 16 drops of water and 4 drops of 1 M NH_4NO_3. **Discard only the wash**
C2. Add 15 drops of 6 M HNO_3 and heat in a water bath until all the solids dissolve. Stirr occassionally.
C3. Discard the solid and transfer the liquid into an evaporating dish.
C4. Add 4 drops of 6M H_2SO_4 stir and evaporate the solution just until the white fumes of SO_3 appears.(!)
C5. Cool and add 1 mL of water. Warm and stir for 3 minutes. Separate solid.

! The evaporation must be done in **a hood.** Corrossive fumes of nitric acid and oxides of nitrogen will be generated, and are extremely toxic.

$PbSO_4$

C6. Add 10 drops of 1M NH_4CH_3COO, to the precipitate warm while constantly stirring for 2 minutes.
C7. Add 2 drops of 0.5 M K_2CrO_4 to the solution. A yellow precipitate confirms Pb^{2+}

Bi^{3+} Cu^{2+} Cd^{2+}

Save this & go to step **D**

Bi^{3+} Cu^{2+} Cd^{2+}

D1. Add 15 M $NH_{3(aq)}$ drop wise until the solution is just basic to litmus. Then add 5 more drops. Stir well. (Read Note 2)

D2. If a white precipitate forms, presence of Bi^{3+} is indicated. Separate it. *

Note 2: $Bi(OH)_3$ is not soluble in excess aqueous ammonia, but Cu^{2+} and Cd^{2+} dissolve by formation of tetrammine salts.

$Bi(OH)_3$

D3. Prepare **FRESH** sodium stannite solution by dropwise adding 4 M NaOH solution to 2 drops of 1 M solution of $SnCl_2$ until a white precipitate forms and then redissolves.

D4. To the white precipitate of $Bi(OH)_3$ add the above freshly prepared sodium stannite solution. Stir throughly. The ***immidiate*** formation of a black solid, which is metallic bismuth confirms Bi^{3+}. (Read Note 3)

Note 3: If lead has come in up to this point, it will form a grey solid with sodium stannite. Bismuth will form a black solid immidiately.

$[Cu(NH_3)_4]^{2+}$, $[Cd(NH_3)_4]^{2+}$

E1. Take 10 drops of the solution in a test tube.

E2. Add 6 M CH_3COOH until acidic.

E3. Add 3 drops of 0.3 M $K_4Fe(CN)_6$.

A deep rose colored precipitate $Cu_2Fe(CN)_6$ confirms Cu^{2+}

A white colored precipitate $Cd_2Fe(CN)_6$ confirms Cd^{2+}

E4. To the remainder solution add a pinch of $Na_2S_2O_4$. Stir well and heat for 3 minutes.

E5. Centrifuge and separate metallic copper.

E6. To the clear solution add 3 drops of thioacetamide, stir well, and heat the mixture for 3 minutes. A yellow precipitate confirms Cd^{2+}

Here you will use the solution in the test tube lebeled GR2B.

$[HgS_2]^{2-}$ $[As(OH)_4]^-$ $[AsS_3]^{3-}$ $[Sb(OH)_4]^-$ $[SbS_3]^{3-}$ $[Sn(OH)_6]^{2-}$ $[SnS_3]^{2-}$

F1. Add 6 M HCl, with stirring, until the solution is just acidic.

F2. Heat the mixture for 5 minutes in a water bath.

F3. Separate, **save the precipitate** and discard the solution.

HgS As_2S_3 Sb_2S_3 SnS_2

G1. Add 1 mL of 6 M HCl to the precipitate, stir well, and heat in a hot water bath for 3 minutes.

G2. Centrifuge and separate the liquid *as soon as possible.*

G3. Add 1 mL of 6 M HCl to the precipitate, stir, and heat the mixture for 3 minutes.

G4 Centrifuge and separate the liquid *as soon as possible,* and combine the two centrifugates. Mark this as ***GR2K*** and place it in the water bath.

HgS As_2S_3

$[SbCl_4]^-$ $[SnCl_6]^{2-}$ (***GR2K***)

HgS As_2S_3

H1. Add 10 drops of 6 M $NH_{3(aq)}$, 5 drops of water, and 10 drops of 6% H_2O_2.
H2. Stir for 1 minute, heat for 5 minutes in a water bath with constant stirring.
H3. Centrifuge and separate the precipitate.

HgS

J1. Add 6 drops of 5% NaClO and 2 drops of 6 M HCl.
J2. Stir, heat for 3 minutes, add 1 mL of water.
J3. Centrifuge and separate

Sulfur (Discard)

J4. Add 4 drops of 1 M SnCl2.
A gery-to-dark grey precipitate confirms Hg^{2+}

AsO_4^{3-}

I1. Wrap a piece of cotton around the tip of the capillary pipet and draw the solution. Transfer the liquid into a clean test tube.
I2. Add 2 drops of 15 M $NH_{3(aq)}$ and 5 drops of magnesia mixture.
A white precipitate confirms AsO_4^{3-} ion

Here you will use the solution in the test tube lebeled GR2K.

$[SbCl_4]^-$ $[SnCl_6]^{2-}$

K1. Add 4 drops of 6 M HCl to the solution.
K2. Place a clean small iron nail and heat the solution for 5 minutes.
K3. Black flakes of Sb will appear. remove the liquid to a clean test tube

Sb

K6. Remove the nail
K7. Add 1 drop of 6 M HNO_3 and a pea head size of $H_2C_2O_4.2H_2O$.
K8. Warm the mixture until the black flake dissolves.
K9. If the flakes do not dissolve add 1 more drop of 6 M HNO_3 and warm the solution.
K10. Add 10 drops of water and 5 drops of thioacetamide to the solution.
K11. Stir throughly and place the test tube in a hot water bath for 3 minutes.
An orange-red precipitate confirms the presence of antimony.

$[SnCl_6]^{2-}$

K4. Add 2 drops of 6 M HCl to the solution and heat for 3 minutes in the water bath.
K5. Add 5 drops of 0.2 M $HgCl_2$ to the warm solution.
Formation of a white grey precipitate confirms tin

ALL CHEMICAL WASTES MUST BE PLACED IN DESIGNATED CONTAINERS FOR PROPER DISPOSAL.

EXPERIMENT 22 QUALITATIVE ANALYSIS - GROUP II
DATA AND LABORATORY REPORT SHEET

Name ______________________Assigned #____ Date ________ Score______

UnKnown #_________

1. Ions Present in the unknown (Group II) are ______________________________

2. Record the color of the precipitate after
   - step A7: ______________
   - step B3: ______________
   - step C5: ______________
   - step D2: ______________
   - step D4: ______________
   - step E3: ______________
   - step E6: ______________
   - step F3: ______________
   - step G4: ______________
   - step H3: ______________
   - step I2: ______________
   - step J4: ______________

3. Refer to steps C1-C4. Why is it necessary to remove HNO_3 before Pb is precipitated as $PbSO_4$?

4. Why fresh sodium stannite must be used for confirmatory test for bismuth(III) ion?

5. To an unknown solution 6 M HCl was added until the solution was acidic to litmus. A Yellow precipitate (A) was obtained when this acidic solution was heated with 1 mL of thioacetamide solution. The precipitate dissolved in 6 M HNO_3. To the resulting solution 6 M H_2SO_4 was added and the mixture was heated until white fumes evolved. The solution was cooled and diluted. 15 M $NH_{3(aq)}$ was added until the solution was basic to litmis. then the solution was acidified by adding CH_3COOH. To this acidified solution when $K_4[Fe(CN)_6]$ solution was added a white precipitate (B) was obtained. Identify A & B.

Pre-Laboratory Assignment Experiment 22
Qualitative Analysis - Group II
Name ________________________________ #________ Score_____

1. Describe the role of thioacetamide in group II analysis.

2. Which of the following is most insoluble? Find its molar solubility.

   PbS, HgS, SnS_2

3. Why would you add HNO_3 and HCl to the solution before separating Group II cations as sulfides?

4. Write the net ionic equations for the confirmatory tests for all the ions in group II.

5. Calculate the S^{2-} concentration in saturated H_2S solution of (a) pH = 1.01 (b) pH = 2.02 and (c) pH = 3.03

FLOW CHART - GROUP III ANALYSIS

Co^{2+} Ni^{2+} Fe^{3+} Mn^{2+} Al^{3+} Cr^{3+} Zn^{2+}

↓

A1. For centrifugate from Gr II add 8 drops of 5% thioacetamide solution and heat the solution for 5 minutes

else

If the solution has only Gr III ions then add 2 drops of 6 M HCl to 10 drops of the solution, dilute to 1 mL, add 8 drops of thioacetamide solution and heat in hot water bath for 5 minutes.

↓

A2. Add 5 drops of 15 M $NH_{3(aq)}$ and stir well, and if the solution is not basic then add dropwise $NH_{3(aq)}$ until basic.

A3. Heat the solution for 5 minutes with frequent stirring.

A4. Separate the precipitate

A5. Wash the precipitate with 10 drops of water and 1 drop of 3 M $NH_{3(aq)}$

↓ Precipitate: CoS NiS FeS MnS $Al(OH)_3$ $Cr(OH)_3$ ZnS

↓ Centrifugate: Gr IV & V

From the precipitate (CoS NiS FeS MnS $Al(OH)_3$ $Cr(OH)_3$ ZnS):

B1. Add 6 drops of water followed by 3 drops of 6 M HCl, stir well for 2 minutes

B2. Separate the mixture *immidiately.*

B3. Repeat steps B1 & B2 , and combine both the centrifugate.

↓ Precipitate: CoS NiS

↓ Centrifugate: Fe^{3+} Mn^{2+} Al^{3+} Cr^{3+} Zn^{2+} — Save the centrifugate lebeled GR3F

From the precipitate (CoS NiS):

C1. Add 8 drops of 6 M HNO_3, stir well, & heat until black residue dissolve. Continue to heat for 3 more minutes with regular stirring.

C2. Add 10 drops of water and make the solution just basic by adding 3 M $NH_{3(aq)}$.

C3. Centrifuge and discard any precipitate that remains (sulfur).

↓

D1. To 6 drop of the above solution add 3 drops of 3 M dimethylglyoxime. Formation of a **cherry-red** precipitate confirms Nickel.

↓

E1. Carefully acidify the remainder of the solution with 6 M HCl.

E2. Add 5 drops of 4 M NH_4SCN soln.

E3. Double the volume by adding acetone.

E4. With a clean finger on top flip the test tube.

If the solution will turn blue or blue-green cobalt is confirmed.

Fe^{3+} Mn^{2+} Al^{3+} Cr^{3+} Zn^{2+} **GR3F**

F1. Add 5 drops of 16 M HNO_3 . Evaporate to a moist residue.
F2. Cool, add 10 drops of water, mix well, and then add 4 M NaOH dropwise, with constant stirring until a precipitate forms.
F3. Then add 10 more drops of 4 M NaOH, stir, and add 10 drops of 6% H_2O_2.
F4. Heat in a water bath for 5 minutes, stirring frequently.
F5. Separate the precipitate.
F6. Wash the precipitate with 1 drop of 4 M NaOH and 10 drops of water.

$Fe(OH)_3$ $Mn(OH)_3$ MnO_2

G1. Add 8 drops of 6 M HNO_3 and 2 drops of 1 M $NaNO_2$
G2. Heat the mixture in a water bath for 6 minutes, stirring frequently.
G3. Cool the test tube under water.
G4. Add 10 drops of water and stir.

H1. Take 10 drops of the solution in to a clean test tube and add 3 drops of 4 M NH_4SCN
A dark red color cinfirms the presence of iron in the solutiuon.

I1. To the reaimder of the solution add 2 drops of 6 M HNO_3 and solid $NaBiO_3$ equivalent to the size of a match head. Stir throughly. Centrifuge. A pink or purple color confirms Manganese.

$[Al(OH)_4]^-$ CrO_4^{2-} $[Zn(OH)_4]^{2-}$

J1. Add 6 M HCl until just acidic and then add 2 more drops.
J2. Centrifuge and discard if there is any precipitate.
J3. Add 6 M $NH_{3(aq)}$ until basic and then add 8 drops. Stir for 1 minute.
J4. A gelateneous precipitate indicates aluminum. Centrifuge and separate the precipitate.

J5. Dissolve the precipitate in 2 drops of 6 M HCl.
J6. Add 5 drops of water and 2 drops of aluminum reagent, followed by enough 6 M $NH_{3(aq)}$ to make the solution basic
An orange to red precipitate confirms aluminum.

K1. Divide the solution to two equal portions.

K4. To the other half of the solution add 5 drops of 5% $(NH_4)_2S$. Formation of a white precipitate , ZnS which is soluble in 1.8 M HCl confirms zinc.

K2. To one portion add 6 M CH_3COOH until acidic.
K3. Add 2 drops of 0.1 M $Pb(CH_3COO)_2$. The presence of chromium is confirmed by the formation of a yellow precipitate.

ALL CHEMICAL WASTES MUST BE PLACED IN DESIGNATED CONTAINERS FOR PROPER DISPOSAL.

EXPERIMENT 22 QUALITATIVE ANALYSIS - GROUP III
DATA AND LABORATORY REPORT SHEET

Name ______________________Assigned #___ Date ________ Score_______
Unknown #__________

1. Ions Present in the unknown (Group I) are ______________________________

2. How Co^{2+} and Ni^{2+} ions are separated from the rest of the group III ions? Describe the chemical principle involved?

3. Why Group III ions are removed in a basic medium?

4. Refer to step B2. Why separate the mixture immidiately?

5. Refer to step F3. What is the role of H_2O_2?

6. Refer to step F4. What is the gelatenous precipitate?

Pre-Laboratory Assignment Experiment 22
Qualitative Analysis - Group III
Name ________________________________ #________ Score_____

1. Why Cr^{3+} and Al^{3+} precipitates as hydroxides rather than sulfides in group III precipitation?

2. Which of the group III cation will
a) give a green solution?
b) dissolve in excess $NH_{3(aq)}$
c) give a pink solution and a black sulfide.

3. Write the net ionic equation for all the confirmatory tests in group III .

APPENDIX
A

GRAPHICAL PRESENTATION OF DATA

The graphing and interpretation of scientific data is very important. Therefore we have included this section in the art of graphing. The construction of a graph is presented.

A graph has designated **x** (**abscissa**) and **y** (**ordinate**) axes. The **x-axis** is the **horizontal axis**, while the **y-axis** is the **vertical axis.**

To plot, first determine which are the x and y data points. Determine the range of values on the x axis and adjust the scale accordingly. The scale should be commensurate with the range of values. Select an appropriate scale for the y axis commensurate with the range of values to be plotted. Each axis should be properly labeled. The data is plotted by placing a small dot for the x and y value followed by a small circle or other symbol around the dot. The circle should represent the error in the measurement. The graph should have a title. In plotting one variable versus another, the variable after the word "versus", is plotted on the **x-axis**, and the other is on the **y-axis.** The **dependent** variable is plotted on the y-axis while the **independent** variable is plotted on the x-axis, e.g. if we say that volume of a gas is **dependent** upon the temperature, then volume is the dependent variable and is plotted on the y-axis, and temperature would be on the x-axis.

A graph does not have to be at **zero** at the origin. Many plots do not go through the origin. Significant figure estimation should be possible using the scale of values. When extrapolation is to be done, the graph must be plotted to allow for extrapolation.

When plotting more than one graph using the same graph paper, it is wise to use a different symbol of the data points, or color.

A smooth curve should be drawn through the data points. The line does not have to pass through every data point. However, it should be close to as many points as possible. Curvature in a plot should be obvious from the locations of the data points. It should be obvious that a curve has been formed.

A typical plot is shown below:

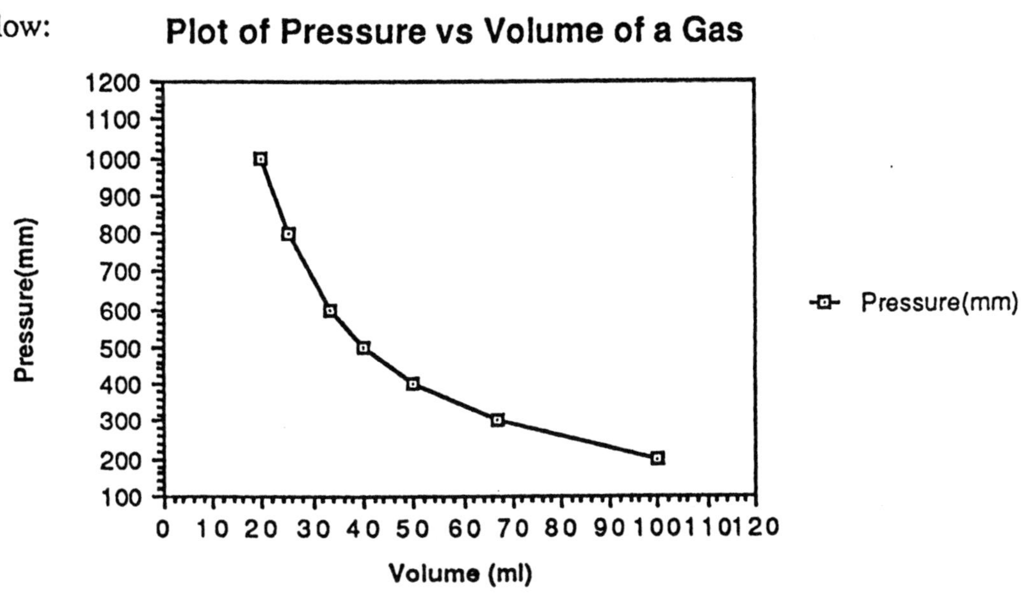

The above plot indicates that there is an inverse relation between the two variables.

Some data do not yield curves, but straight lines. Again, it should be apparent from the location of the points, that a straight line is in order. These are linear relationships and the line which corresponds to the best fit of the data can readily be ascertained by simply drawing a straight line with a ruler. The equation of a straight line is **$y = mx + b$, m = slope, and b is the y intercept**, the point at which the plot crosses the y-axis. The slope of the line is the **rise** $(y_2 - y_1)$ divided by the **run** $(x_2 - x_1)$.

$m = y_2 - y_1 / x_2 - x_1$ The slope is determined by selecting two points on the line and identifying them as (x_1, y_1) and (x_2, y_2), where the 1 and 2 refer to the first and second data points respectively.

A typical straight line plot is shown below:

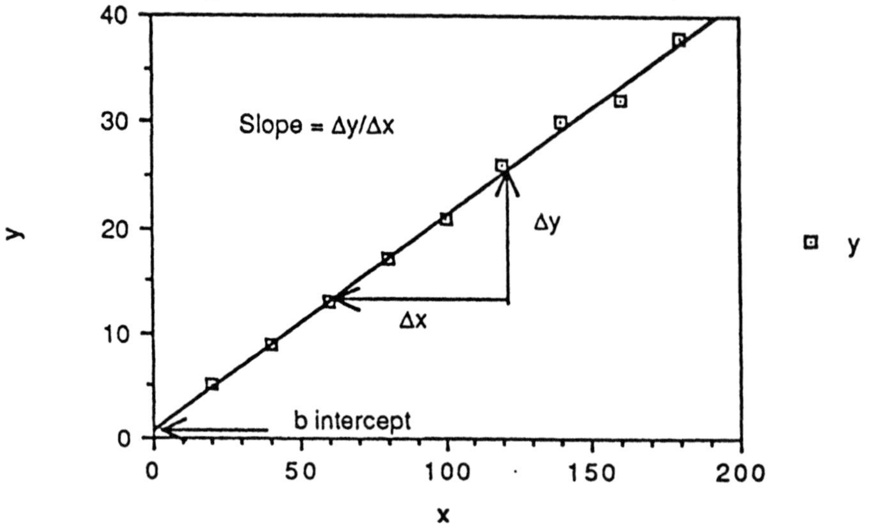

The above plot indicates a **direct relation** existing between the variables. As one variable increases, the other increases.

The slope of the plot would be a (+) slope.

The curve below indicates an **inverse relation** among the variables where the slope would be (-). As one variable increases, the other variable decreases.

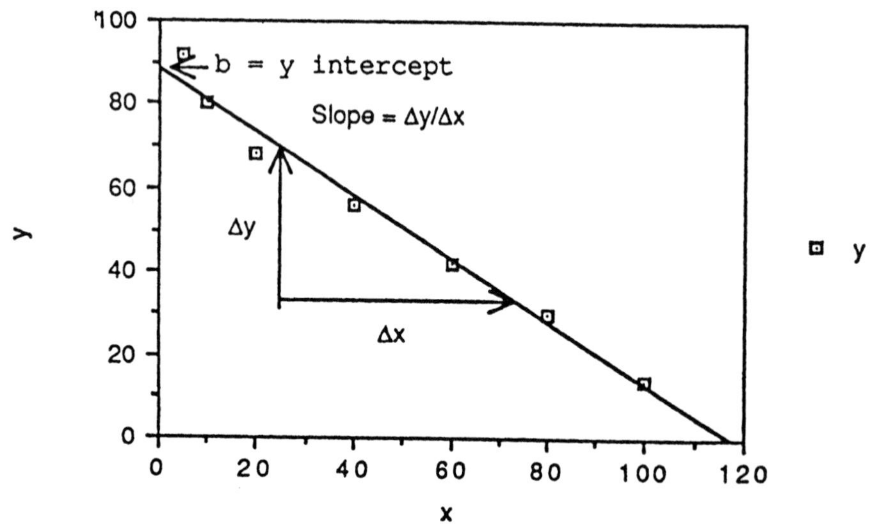

The computer can be used to plot data. If the data is to be plotted by computer, there are a number of different plotting software packages. If you are allowed to use one , please learn how to use it. Often times gridded plots are needed just as is graph paper. Therefore, if you are using the computer, you must be able to grid it properly for slope determinations. The above plot with a reasonable grid is shown below.

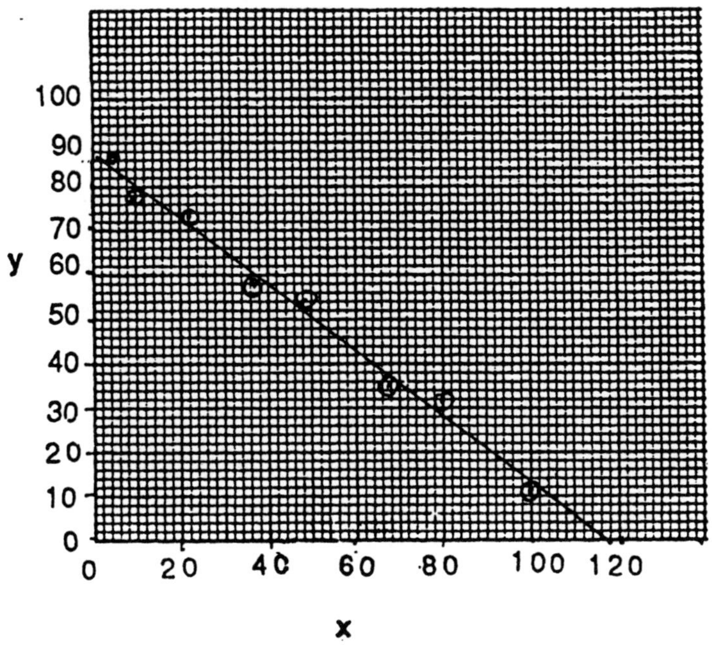

When plotting negative numbers, the number line should be recalled. As in the positive number plots, negative numbers are plotted according to the number scale. Be aware of the position of negative numbers.

4
3
2
1
-4 -3 -2 -1 0 1 2 3 4
-1
-2
-3
-4

A typical plot of data with negative y-axis points is shown below. if negative x-axis were plotted, the (-) numbers would go to the left of zero, -4 -3 -2 -1 0 .

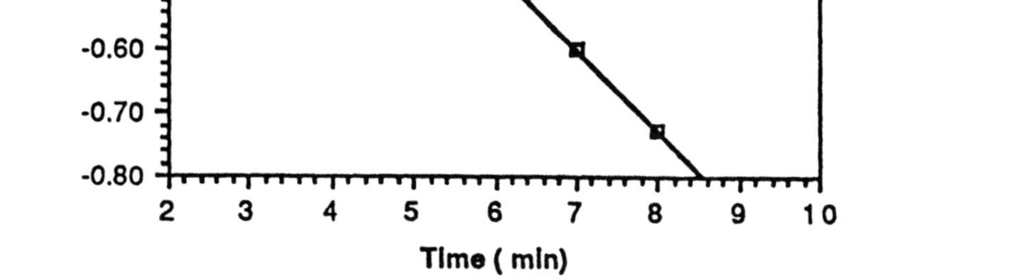

The graphing of data requires care and concern. Plots should appear legible and neat.

APPENDIX
B

VAPOR PRESSURE OF WATER
TEMPERATURE VARIATION

Temperature(°C)	Pressure (torr)	Temperature(°C)	Pressure (torr)
0	4.6	26	25.2
1	4.9	27	26.7
2	5.3	28	28.3
3	5.7	29	30.0
4	6.1	30	31.8
5	6.5	31	33.7
6	7.0	32	35.7
7	7.5	33	37.7
8	8.0	34	39.9
9	8.6	35	42.2
10	9.2	40	55.3
11	9.8	45	71.9
12	10.5	50	92.5
13	11.2	55	118.0
14	12.0	60	149.4
15	12.8	65	187.5
16	13.6	70	233.7
17	14.5	75	289.1
18	15.5	80	355.1
19	16.5	85	433.6
20	17.5	90	525.8
21	18.7	95	633.9
22	19.8	97	682.1
23	21.1	99	733.2
24	22.4	100	760.0
25	23.8	101	787.6

DENSITY OF WATER AT VARIOUS TEMPERATURES

Temp. (°C)	Density (g/cm^3)	Temp. (°C)	Density (g/cm^3)
-5 (ice)	0.99930	25	0.99707
0	0.99987	26	0.99681
4	1.00000	27	0.99654
5	0.99999	28	0.99626
10	0.99973	29	0.99597
15	0.99913	30	0.99567
20	0.99823	31	0.99537
21	0.99802	40	0.99224
22	0.99780	50	0.98807
23	0.99756	60	0.98324
24	0.99732	100	0.95838

APPENDIX C
WATER SOLUBILITY AND Ksp

GENERAL SOLUBILITY RULES FOR COMMON SALTS IN WATER

NO_3^-	Most nitrates are **soluble**
Na^+, K^+, NH_4^+	Most sodium, potassium and ammonium salts are **soluble**
$C_2H_3O_2^-$	Most acetates are **soluble** except silver acetate
Cl^-	Most chlorides are **soluble** except $PbCl_2$, AgCl and Hg_2Cl_2
SO_4^{-2}	Most sulfates are **soluble** except $BaSO_4$, $PbSO_4$
CO_3^{-2} and PO_4^{-3}	Most carbonates and phosphates are **insoluble** except those of Na^+, K^+, NH_4^+
OH^-	Most hydroxides are **insoluble,** except NaOH, KOH, $Ba(OH)_2$. $Ca(OH)_2$ is only slightly soluble.
S^{-2}	Most sulfides are **insoluble** except those of Na^+, K^+, NH_4^+ and Mg^{+2}, Ca^{+2}, Sr^{+2}, and Ba^{+2}
Ag^+	Most silver salts are **insoluble** except $AgNO_3$ and $AgClO_3$

SOLUBILITY PRODUCT CONSTANTS (18-25°C)

Compound	Ksp	Compound	Ksp
ACETATES			
$AgC_2H_3O_2$	4×10^{-3}		
HALIDES & CYANIDES			
AgCN	1×10^{-16}	CuCl	3.2×10^{-7}
AgCl	1.8×10^{-10}	Hg_2Cl_2	1.1×10^{-18}
AgBr	5×10^{-13}	$PbCl_2$	1.6×10^{-5}
AgI	8.5×10^{-17}	CaF_2	1.7×10^{-10}

Ksp Continued

Compound	Ksp	Compound	Ksp
CARBONATES			
Ag_2CO_3	8×10^{-12}	$BaCO_3$	1.6×10^{-9}
$CaCO_3$	4.8×10^{-9}	$CuCO_3$	2.5×10^{-10}
$FeCO_3$	2×10^{-11}	$PbCO_3$	1.5×10^{-13}
CHROMATES			
Ag_2CrO_4	2×10^{-12}	$BaCrO_4$	1.2×10^{-10}
$PbCrO_4$	2×10^{-16}	$SrCrO_4$	3.6×10^{-5}
HYDROXIDES			
$Al(OH)_3$	10^{-33}	$Ca(OH)_2$	1.3×10^{-6}
$Cr(OH)_3$	10^{-30}	$Cu(OH)_2$	2×10^{-19}
$Fe(OH)_2$	2×10^{-15}	$Fe(OH)_3$	10^{-37}
$Mg(OH)_2$	9×10^{-12}	$Pb(OH)_2$	4×10^{-15}
$Sn(OH)_2$	10^{-27}	$Zn(OH)_2$	5×10^{-17}
SULFATES			
Ag_2SO_4	1.7×10^{-5}	$BaSO_4$	1.5×10^{-9}
$CaSO_4$	2.4×10^{-5}	$PbSO_4$	1.3×10^{-8}
SULFIDES			
Ag_2S	10^{-50}	CdS	2×10^{-28}
CuS	6×10^{-36}	FeS	6×10^{-18}
HgS	10^{-50}	MnS	3×10^{-13}
NiS	10^{-22}	PbS	7×10^{-28}
SnS	10^{-27}	ZnS	4.5×10^{-24}
OXALATES			
BaC_2O_4	1.5×10^{-8}	CaC_2O_4	1.3×10^{-9}
MgC_2O_4	8.6×10^{-5}	SrC_2O_4	5.6×10^{-8}

APPENDIX
D

DATA TREATMENT

In carrying out measurements, it is not possible to eliminate all the uncertainty in analyses. Every measurement is subject to uncertainties. Since all of the uncertainties cannot be ever eliminated, the true value of any measurement will remain unknown. If the value of the error in a measurement can be determined, then we can place limits (a range of values) that the true value of a measured quantity can have. Of utmost importance is the **accuracy of the data, how near is the data to the true value.** To obtain accurate results, the measurement devices themselves must be calibrated using known standards. However, even using such accurate devices do not eliminate all of the errors. These errors are called **systematic errors.** Repitition of the same measurement can yield different values. Systematic errors are unavoidable and occur as a result of the differences in the ability to observe, inherent in individuals. Thus all experimental measurements are subject to systematic errors. As a result of this, chemists and other scientists are concerned about the **precision** of scientific data, **how near are the measurements to each other.** The measurements must be conducted under the exact same conditions. Errors of this kind are called **random errors.** Random errors which are small are indicative of precise data (high precision). Large random errors result in very low precision, the measurements are far apart from each other. In general, data with high precision is also of high accuracy.

If a large number of measurements are made , there would be a fluctuation about a value called the **mean, or average, x** . The average or mean is determined by taking the sum of the mesurements and dividing by the number of measurements, N , $x_i = (x1 + x2 + x3 + x4)/N$, N = 4 here. Precision can be expressed in several ways. Statistically, the **Standard Deviation, s,** is used as a measure of the precision of measurements even when the number of measurements is small. To compute the standard deviation, we first determine the deviation of each measurement from the mean, **dev. 1 = (x1- x) , dev. 2 = (x2-x)** , etc. This is repeated for each data point. The deviations must then be squared , **$(x1-x)^2$** , etc. The deviations are then summed and divided by the number of points, N degrees of freedom minus one, (N-1) . The square root is then taken. The formula below is for the standard deviation.

$$s = \sqrt{\frac{\sum (x_i - x)^2}{N - 1}}$$

The value of the measurement is reported as the mean ± standard deviation, x ± s. A large value of s is indicative of low precision, while a small value of s indicates that the precision is high. Therefore, s is a statistical measure of the precision of the experimental data. The standard deviation means that 68% of the subsequent values lie within **± s.** If in a series of measurements, the mean was 10.5 and the standard deviation = 0.67, the vmeasurement would be reported as 10.5±0.67 , which means that 68% of all subsequent values would lie in the range 10.5 ± 0.67.

The **percent error** is used as a measure of the **accuracy** of a measurement, scientist use the formula :

$$\% \text{ error} = \frac{|x-y|}{y} \times 100,$$

The absolute value of the difference between your experimental value x and the "accepted value" x 100 divided by the accepted value.

The Chemical Rubber Company's Handbook of Chemistry and Physics , and other reference texts are used in the determination of the " **accepted values**" for various measurements.

APPENDIX
E

ACID BASE CONCENTRATIONS

Reagent	Molarity	%	Density (g/ml)	Formula	Required conc. Reagent / liter
Acetic acid, Glacial	17M	99.5	1.05	CH_3COOH	
Acetic acid	6M (dilute)			345ml	
Acetic acid	3M				172ml
Acetic acid	1M				57.5ml
Hydrochloric acid, conc.	12.0M	36	1.18	HCl	
Hydrochloric acid	6.0M (dilute)				496ml
Hydrochloric acid	3.0M				248ml
Hydrochloric acid	1.0M				82.6ml
Nitric acid, conc.	15.7M	72	1.42	HNO_3	
Nitric acid	6.0M (dilute)				382ml
Nitric acid	4.0M				255ml
Nitric acid	1.0M				63.7ml
Sulfuric acid, conc.	17.8M	96	1.84	H_2SO_4	
Sulfuric acid	6.0M(dilute)				337ml
Sulfuric acid	3.0M(dilute)				169ml
Sulfuric acid	1.0M				56.2ml
Phosphoric acid, conc.	14.7M	85	1.87	H_3PO_4	
Phosphoric acid	6.0M (dilute)				408ml
Phosphoric acid	1.0M				68.0ml
Phosphoric acid	0.1M				6.8ml
Ammonia , conc.	14.8M	58	0.90	$NH_3(aq)$	
Ammonia	6.0M (dilute)				405ml
Ammonia	3.0M				203ml
Ammonia	1.0M				67.6ml

APPENDIX
F
FUNDAMENTAL CONSTANTS AND CONVERSION FACTORS

CONSTANTS

	Symbol	Magnitude			
Speed of Light	(c)	2.9998E8 m/s	186,272 miles/s		
Avagadro's Number	N	6.022E23 atoms/mol ,	6.022E23 molecules/mol		
Gas Constant	R	1.986 cal / K.mol =	0.08205 l-atm / K.mol =	62.36 L-torr / K.mol	
	R	8.314 J / K. mol			
Faraday Constant	F	96,487 C/mol e^-			
Planck's Constant	h	6.626196 E-27 erg-s	6.626196E-34 J-s		
Boltzman Constant	k	1.38054E- 23 J / K			
Pi,	π	3.14159			
Natural log base,	e	2.71828			
ln x		2.3026 log x			
1Atomic mass unit	amu	1.6606E-24 g			
Electron Rest Mass	m_e	9.1095E-31 kg			
Proton Rest Mass	m_p	1.6726E-27 kg			
Molar Volume at STP	V_m	22.414 L / mol			
Elementary Charge	e	1.6022E-19 C			
Acceleration of gravity	g	9.806m/s^2			
Rydberg Constant for H	R_H	109,678cm^{-1}	1.09678E7m^{-1}	3.289E15s^{-1}	2.1799E-18J

LENGTH

1 inch (in) = 2.54 cm   1mile = 1.609km   1 Å = 1E-8 cm = 1 E-10 m   1 Yard= 0.9144 m
1 km = 1000 m   1decimeter= 0.1 m   1 centimeter = 0.01 m   1 millimeter = 0.001 m
1 m= 39.37 in = 1.094 yards

VOLUME

1 Liter = 1.057 qts   0.9463Liter (L)= 1quart   1milliliter = 1 cubic centimeter (cc)   1 Liter = 1000 cc = 61.2 in^3
1 fluid ounce= 29.57 ml

MASS

1 lb = 453.6 g   1 kg = 1000 g   1 Metric ton = 1E6 g = 1000 kg   1 Long ton = 2240 lb
1 kg = 2.205 lb   1 Short ton = 2000 lb = 907.185 kg   1 Ounce = 28.35 g = 437.5 grains

ENERGY

1 calorie = 4.184 J   1 Joule = 9.48E- 4 Btu   1Liter-atmosphere= 101.32 J   1 Electron volt ev = 23.06 kcal / mol
1 ev = 1.60 E-19 J   1 J = 1E7 erg   1 kw hr = 8.606 E5 cal = 3.601E6J

FORCE

1 Newton N = 1kg.m/s^2

PRESSURE

1 Torr = 1 mm Hg   1 atmosphere, atm = 760 torr = 760 mm Hg = 14.7 psi = 101,325 N / m^2 =
101.325 Pascals , Pa   1 Pascal = kg/m.s^2 = N / m^2 1 atm = 29.92 in Hg
1 torr = 1 mm = 133.3 N / m^2

TEMPERATURE

$^oK = 273.15 + {^oC}$ $0^oK = -273.15\ {^oC}$ (Celsius) $^oC = 5/9\ (^oF - 32)$ $^oF = 9/5\ {^oC} + 32$

APPENDIX
G
COLOR CHANGES AND pH OF INDICATORS

Indicator	pH Range	Color Change	Solvent system
Phenol Red	6.8-8.2	Yellow to Red	($NaOH$&H_2O)
Ethyl Red	4.0-5.8	Colorless to Red	($MeOH$&H_2O)
Methyl Red	4.4-6.2	Red to Yellow	($EtOH$&H_2O)
Chlorphenol Red	4.8-6.8	Yellow to Red	($NaOH$&H_2O)
Congo Red	3.0-5.0	Blue to Red	(H_2O)
Neutral Red	6.8-8.0	Red to Amber	($EtOH$&H_2O)
Methyl Violet	0.2-3.0	Yellow-Blue-Violet	(H_2O)
Bromcresol Purple	5.2-6.8	Yellow to Purple	($NaOH$&H_2O)
Thymol Blue	8.0-9.6	Yellow to Blue	($NaOH$&H_2O)
Thymol Blue	1.2-2.8	Red to Yellow	(H_2O)
Bromthymol Blue	6.0-7.6	Yellow to Blue	($NaOH$&H_2O)
Bromphenol Blue	3.0-4.6	Yellow-Blue-Violet	($NaOH$&H_2O)
Thymolphthalein	9.4-10.6	Colorless to Blue	($EtOH$&H_2O)
Methyl Orange	3.1-4.4	Red - Orange-Yellow	(H_2O)
Orange IV	1.3-3.0	Red to Yellow	(H_2O)
Phenolphthalein	8.2-10.0	Colorless to Red	($EtOH$&H_2O)
Bromcresol Green	3.8-5.4	Yellow to Blue	($NaOH$&H_2O)
Alizarin Yellow R	10.0-12.0	Yellow to Red	($EtOH$&H_2O)
Indigo Carmine	11.4-13.0	Blue to Yellow	($EtOH$&H_2O)

APPENDIX H
LOGARITHM TABLE

Four-Place Table of Logarithms

No.	0	1	2	3	4	5	6	7	8	9	1	2	3	4	5	6	7	8	9
10	0000	0043	0086	0128	0170	0212	0253	0294	0334	0374	4	8	12	17	21	25	29	33	37
11	0414	0453	0492	0531	0569	0607	0645	0682	0719	0755	4	8	11	15	19	23	26	30	34
12	0792	0828	0864	0899	0934	0969	1004	1038	1072	1106	3	7	10	14	17	21	24	28	31
13	1139	1173	1206	1239	1271	1303	1335	1367	1399	1430	3	6	10	13	16	19	23	26	29
14	1461	1492	1523	1553	1584	1614	1644	1673	1703	1732	3	6	9	12	15	18	21	24	27
15	1761	1790	1818	1847	1875	1903	1931	1959	1987	2014	3	6	8	11	14	17	20	22	25
16	2041	2068	2095	2122	2148	2175	2201	2227	2253	2279	3	5	8	11	13	16	18	21	24
17	2304	2330	2355	2380	2405	2430	2455	2480	2504	2529	2	5	7	10	12	15	17	20	22
18	2553	2577	2601	2625	2648	2672	2695	2718	2742	2765	2	5	7	9	12	14	16	19	21
19	2788	2810	2833	2856	2878	2900	2923	2945	2967	2989	2	4	7	9	11	13	16	18	20
20	3010	3032	3054	3075	3096	3118	3139	3160	3181	3201	2	4	6	8	11	13	15	17	19
21	3222	3243	3263	3284	3304	3324	3345	3365	3385	3404	2	4	6	8	10	12	14	16	18
22	3424	3444	3464	3483	3502	3522	3541	3560	3579	3598	2	4	6	8	10	12	14	15	17
23	3617	3636	3655	3674	3692	3711	3729	3747	3766	3784	2	4	6	7	9	11	13	15	17
24	3802	3820	3838	3856	3874	3892	3909	3927	3945	3962	2	4	5	7	9	11	12	14	16
25	3979	3997	4014	4031	4048	4065	4082	4099	4116	4133	2	3	5	7	9	10	12	14	15
26	4150	4166	4183	4200	4216	4232	4249	4265	4281	4298	2	3	5	7	8	10	11	13	15
27	4314	4330	4346	4362	4378	4393	4409	4425	4440	4456	2	3	5	6	8	9	11	13	14
28	4472	4487	4502	4518	4533	4548	4564	4579	4594	4609	2	3	5	6	8	9	11	12	14
29	4624	4639	4654	4669	4683	4698	4713	4728	4742	4757	1	3	4	6	7	9	10	12	13
30	4771	4786	4800	4814	4829	4843	4857	4871	4886	4900	1	3	4	6	7	9	10	11	13
31	4914	4928	4942	4955	4969	4983	4997	5011	5024	5038	1	3	4	6	7	8	10	11	12
32	5051	5065	5079	5092	5105	5119	5132	5145	5159	5172	1	3	4	5	7	8	9	11	12
33	5185	5198	5211	5224	5237	5250	5263	5276	5289	5302	1	3	4	5	6	8	9	10	12
34	5313	5328	5340	5353	5366	5378	5391	5403	5416	5428	1	3	4	5	6	8	9	10	11
35	5441	5453	5465	5478	5490	5502	5514	5527	5539	5551	1	2	4	5	6	7	9	10	11
36	5563	5575	5587	5599	5611	5623	5635	5647	5658	5670	1	2	4	5	6	7	8	10	11
37	5682	5694	5705	5717	5729	5740	5752	5763	5775	5786	1	2	3	5	6	7	8	9	10
38	5798	5809	5821	5832	5843	5855	5866	5877	5888	5899	1	2	3	5	6	7	8	9	10
39	5911	5922	5933	5944	5955	5966	5977	5988	5999	6010	1	2	3	4	5	7	8	9	10
40	6021	6031	6042	6053	6064	6075	6085	6096	6107	6117	1	2	3	4	5	6	8	9	10
41	6128	6138	6149	6160	6170	6180	6191	6201	6212	6222	1	2	3	4	5	6	7	8	9
42	6232	6243	6253	6263	6274	6284	6294	6304	6314	6325	1	2	3	4	5	6	7	8	9
43	6335	6345	6355	6365	6375	6385	6395	6405	6415	6425	1	2	3	4	5	6	7	8	9
44	6435	6444	6454	6464	6474	6484	6493	6503	6513	6522	1	2	3	4	5	6	7	8	9
45	6532	6542	6551	6561	6571	6580	6590	6599	6609	6618	1	2	3	4	5	6	7	8	9
46	6628	6637	6646	6656	6665	6675	6684	6693	6702	6712	1	2	3	4	5	6	7	7	8
47	6721	6730	6739	6749	6758	6767	6776	6785	6794	6803	1	2	3	4	5	5	6	7	8
48	6812	6821	6830	6839	6848	6857	6866	6875	6884	6893	1	2	3	4	4	5	6	7	8
49	6902	6911	6920	6928	6937	6946	6955	6964	6972	6981	1	2	3	4	4	5	6	7	8
50	6990	6998	7007	7016	7024	7033	7042	7050	7059	7067	1	2	3	3	4	5	6	7	8
51	7076	7084	7093	7101	7110	7118	7126	7135	7143	7152	1	2	3	3	4	5	6	7	8
52	7160	7168	7177	7185	7193	7202	7210	7218	7226	7235	1	2	2	3	4	5	6	7	7
53	7243	7251	7259	7267	7275	7284	7292	7300	7308	7316	1	2	2	3	4	5	6	6	7
54	7324	7332	7340	7348	7356	7364	7372	7380	7388	7396	1	2	2	3	4	5	6	6	7
55	7404	7412	7419	7427	7435	7443	7451	7459	7466	7474	1	2	2	3	4	5	5	6	7
56	7482	7490	7497	7505	7513	7520	7528	7536	7543	7551	1	2	2	3	4	5	5	6	7
57	7559	7566	7574	7582	7589	7597	7604	7612	7619	7627	1	2	2	3	4	5	5	6	7
58	7634	7642	7649	7657	7664	7672	7679	7686	7694	7701	1	1	2	3	4	4	5	6	7
59	7709	7716	7723	7731	7738	7745	7752	7760	7767	7774	1	1	2	3	4	4	5	6	7
60	7782	7789	7796	7803	7810	7818	7825	7832	7839	7846	1	1	2	3	4	4	5	6	6
61	7853	7860	7868	7875	7882	7889	7896	7903	7910	7917	1	1	2	3	4	4	5	6	6
62	7924	7931	7938	7945	7952	7959	7966	7973	7980	7987	1	1	2	3	3	4	5	6	6
63	7992	8000	8007	8014	8021	8028	8035	8041	8048	8055	1	1	2	3	3	4	5	5	6
64	8062	8069	8075	8082	8089	8096	8102	8109	8116	8122	1	1	2	3	3	4	5	5	6
65	8129	8136	8142	8149	8156	8162	8169	8176	8182	8189	1	1	2	3	3	4	5	5	6
66	8195	8202	8209	8215	8222	8228	8235	8241	8248	8254	1	1	2	3	3	4	5	5	6
67	8261	8267	8274	8280	8287	8293	8299	8306	8312	8319	1	1	2	3	3	4	5	5	6
68	8325	8331	8338	8344	8351	8357	8363	8370	8376	8382	1	1	2	3	3	4	4	5	6
69	8388	8395	8401	8407	8414	8420	8426	8432	8439	8445	1	1	2	2	3	4	4	5	6
70	8451	8457	8463	8470	8476	8482	8488	8494	8500	8506	1	1	2	2	3	4	4	5	6
71	8513	8519	8525	8531	8537	8543	8549	8555	8561	8567	1	1	2	2	3	4	4	5	5
72	8573	8579	8585	8591	8597	8603	8609	8615	8621	8627	1	1	2	2	3	4	4	5	5
73	8633	8639	8645	8651	8657	8663	8669	8675	8681	8686	1	1	2	2	3	4	4	5	5
74	8692	8698	8704	8710	8716	8722	8727	8733	8739	8745	1	1	2	2	3	4	4	5	5
75	8751	8756	8762	8768	8774	8779	8785	8791	8797	8802	1	1	2	2	3	3	4	5	5
76	8808	8814	8820	8825	8831	8837	8842	8848	8854	8859	1	1	2	2	3	3	4	5	5
77	8865	8871	8876	8882	8887	8893	8899	8904	8910	8915	1	1	2	2	3	3	4	4	5
78	8921	8927	8932	8938	8943	8949	8954	8960	8965	8971	1	1	2	2	3	3	4	4	5
79	8976	8982	8987	8993	8998	9004	9009	9015	9020	9025	1	1	2	2	3	3	4	4	5
80	9031	9036	9042	9047	9053	9058	9063	9069	9074	9079	1	1	2	2	3	3	4	4	5
81	9085	9090	9096	9101	9106	9112	9117	9122	9128	9133	1	1	2	2	3	3	4	4	5
82	9138	9143	9149	9154	9159	9165	9170	9175	9180	9186	1	1	2	2	3	3	4	4	5
83	9191	9196	9201	9206	9212	9217	9222	9227	9232	9238	1	1	2	2	3	3	4	4	5
84	9243	9248	9253	9258	9263	9269	9274	9279	9284	9289	1	1	2	2	3	3	4	4	5
85	9294	9299	9304	9309	9315	9320	9325	9330	9335	9340	1	1	2	2	3	3	4	4	5
86	9345	9350	9355	9360	9365	9370	9375	9380	9385	9390	1	1	2	2	3	3	4	4	5
87	9395	9400	9405	9410	9415	9420	9425	9430	9435	9440	0	1	1	2	2	3	3	4	4
88	9445	9450	9455	9460	9465	9469	9474	9479	9484	9489	0	1	1	2	2	3	3	4	4
89	9494	9499	9504	9509	9513	9518	9523	9528	9533	9538	0	1	1	2	2	3	3	4	4
90	9542	9547	9552	9557	9562	9566	9571	9576	9581	9586	0	1	1	2	2	3	3	4	4
91	9590	9595	9600	9605	9609	9614	9619	9624	9628	9633	0	1	1	2	2	3	3	4	4
92	9638	9643	9647	9652	9657	9661	9666	9671	9675	9680	0	1	1	2	2	3	3	4	4
93	9685	9689	9694	9699	9703	9708	9713	9717	9722	9727	0	1	1	2	2	3	3	4	4
94	9731	9736	9741	9745	9750	9754	9759	9763	9768	9773	0	1	1	2	2	3	3	4	4
95	9777	9782	9786	9791	9795	9800	9805	9809	9814	9818	0	1	1	2	2	3	3	4	4
96	9823	9827	9832	9836	9841	9845	9850	9854	9859	9863	0	1	1	2	2	3	3	4	4
97	9868	9872	9877	9881	9886	9890	9894	9899	9903	9908	0	1	1	2	2	3	3	4	4
98	9912	9917	9921	9926	9930	9934	9939	9943	9948	9952	0	1	1	2	2	3	3	4	4
99	9956	9961	9965	9969	9974	9978	9983	9987	9991	9996	0	1	1	2	2	3	3	3	4
	0	1	2	3	4	5	6	7	8	9	1	2	3	4	5	6	7	8	9

APPENDIX
I
PERIODIC CHART

1 H 1.0079	Atomic number Symbol Atomic mass

I A	II A	III B	IV B	V B	VI B	VII B	VIII B			I B	II B	III A	IV A	V A	VI A	VII A	VIII A
1 H 1.0079																	2 He 4.0026
3 Li 6.94	4 Be 9.0122											5 B 10.81	6 C 12.022	7 N 14.0067	8 O 15.9994	9 F 18.9984	10 Ne 20.179
11 Na 22.989	12 Mg 24.305											13 Al 26.9815	14 Si 28.0855	15 P 30.9737	16 S 32.06	17 Cl 35.453	18 Ar 39.948
19 K 39.098	20 Ca 40.08	21 Sc 44.9559	22 Ti 47.88	23 V 50 9415	24 Cr 51.996	25 Mn 54.9380	26 Fe 55.847	27 Co 58.9332	28 Ni 58.69	29 Cu 63.546	30 Zn 65.38	31 Ga 69.72	32 Ge 72.59	33 As 74.9216	34 Se 78.96	35 Br 79.904	36 Kr 83.80
37 Rb 85.467	38 Sr 87.62	39 Y 88.9059	40 Zr 91.22	41 Nb 92.9064	42 Mo 95.94	43 Tc (98)	44 Ru 101.07	45 Rh 102.9055	46 Pd 106.42	47 Ag 107.868	48 Cd 112.41	49 In 114.82	50 Sn 118.69	51 Sb 121.75	52 Te 127.60	53 I 126.904	54 Xe 131.29
55 Cs 132.905	56 Ba 137.33	57 La * 138.9055	72 Hf 178.49	73 Ta 180.948	74 W 183.85	75 Re 186.207	76 Os 190.2	77 Ir 192.22	78 Pt 195.08	79 Au 196.966	80 Hg 200.59	81 Tl 204.383	82 Pb 207.2	83 Bi 208.9804	84 Po (209)	85 At (210)	86 Rn (222)
87 Fr (223)	88 Ra 226.025	89 Ac† 227.0278	104 Unq§ (261)	105 Unp§ (262)	106 Unh§ (263)												

Lanthanide Series *	58 Ce 140.12	59 Pr 140.907	60 Nd 144.24	61 Pm (145)	62 Sm 150.36	63 Eu 151.96	64 Gd 157.25	65 Tb 158.925	66 Dy 162.50	67 Ho 164.930	68 Er 167.26	69 Tm 168.934	70 Yb 173.04	71 Lu 174.967
Actinide Series †	90 Th 232.038	91 Pa 231.036	92 U 238.029	93 Np 237.048	94 Pu (244)	95 Am (243)	96 Cm (247)	97 Bk (249)	98 Cf (251)	99 Es (254)	100 Fm (257)	101 Md (258)	102 No (259)	103 Lr (260)

§ IUPAC has not adopted official names for these elements

10 DIVISIONS PER CENTIMETER

10 DIVISIONS PER CENTIMETER

10 DIVISIONS PER CENTIMETER

10 DIVISIONS PER CENTIMETER

10 DIVISIONS PER CENTIMETER

10 DIVISIONS PER CENTIMETER

10 DIVISIONS PER CENTIMETER

10 DIVISIONS PER CENTIMETER